Histories of the Dustheap

Urban and Industrial Environments

Series editor: Robert Gottlieb, Henry R. Luce Professor of Urban and Environmental Policy, Occidental College

Histories of the Dustheap

Waste, Material Cultures, Social Justice

edited by Stephanie Foote and Elizabeth Mazzolini

The MIT Press
Cambridge, Massachusetts
London, England

MIT Press books may be purchased at special quantity discounts for business or sales promotional use. For information, please e-mail special_sales@mitpress.mit.edu or write to Special Sales Department, The MIT Press, 55 Hayward Street, Cambridge, MA 02142.

This book was set in Sabon by the MIT Press. Printed on recycled paper and bound in the United States of America.

Library of Congress Cataloging-in-Publication Data

Histories of the dustheap: waste, material cultures, social justice / edited by Stephanie Foote and Elizabeth Mazzolini.
    p. cm. — (Urban and industrial environments)
Includes bibliographical references and index.
ISBN 978-0-262-01799-2 (hardcover : alk. paper) — ISBN 978-0-262-51782-9 (pbk. :alk. paper)
1. Refuse and refuse disposal—Social aspects. 2. Refuse and refuse disposal—Political aspects. 3. Waste products—Social aspects. 4. Waste products—Political aspects. 5. Material culture. I. Foote, Stephanie. II. Mazzolini, Elizabeth, 1973–
HD4482.H57   2012
363.72'8—dc23
2012007130

10 9 8 7 6 5 4 3 2 1

Contents

Acknowledgments

As is well known, books are actually group projects, in spite of the single or few names that appear on their covers. We felt this collaborative spirit especially at work in the crafting of this book, and so we feel grateful to all the contributors, whose superb essays, openness to the collaborative spirit, productive engagements with feedback, and timeliness helped make this book what it is. We are also grateful to the anonymous readers of a draft for their insights, which drew out the book's strengths, and to our editors at the MIT Press, Robert Gottlieb and Clay Morgan, who shepherded the book through its publication.

As an editing and writing team, though, we feel our main debt of gratitude is to each other. Neither of us had coauthored or coedited anything with anyone before embarking on this project, and we each needed help figuring out the collaborative professional relationship. We feel we did this well with each other's assistance, and each of us is impressed with and appreciative of how productive our relationship became over time. Not to belabor the garbage metaphor, but even the text that got thrown out was useful and moved us forward.

Elizabeth would like to thank her parents, Mark and Pat Mazzolini, and her brother, David Mazzolini, for their support. Nonhuman friends Lola, Maebe, Levon, and Francis provide many pleasurable wastes of time. Most thanks (accompanied by love) go to Chad Lavin for his remarkably unwavering encouragement, patience reading multiple drafts, and inevitably useful insights. Baby Walter arrived during the production of this book, and he's the best.

Stephanie would like to thank the remarkable Cris Mayo, who listened to her talk about garbage every day and helped her see what was worth saving in her own writing and thinking. Peter, Deby, Zak, and Krystina Foote also deserve special mention as well as praise for their eternal wit, graciousness, and love.

Introduction: Histories of the Dustheap

Elizabeth Mazzolini and Stephanie Foote

I

In 2007, a flotilla of what are commonly called rubber duckies arrived on Britain's beaches. Like most mass-produced cheap toys, they were not rubber but rather plastic. They, along with their plastic frog, turtle, and beaver companions (approximately twenty-nine thousand toys all told) were produced by the same manufacturer, and had been adrift for fifteen years after a container on the ship carrying them overturned in 1992. Over the years the ducks and their friends visited Hawaii, floated along the coast of Alaska to the Bering Strait, and made their way through the Arctic Ocean to Iceland and New England.

Along their journey, the plastic toys aided oceanographers and other researchers studying surface ocean currents and global weather patterns; they were easy to spot, and more likely to be reported than official floats specifically designed for the task of studying currents. In the scheme of things, the toys' unintended voyage is scarcely more unlikely than the trip they were intended to make from factories in China to toy stores in the United States. Their prolonged ocean voyage was, on the face of it, more "natural" than their originally planned route because it followed ocean currents instead of the socially constructed paths of global commerce. But in some sense, it was precisely this strange palimpsest of natural and unnatural that seemed to draw the attention of the media. The incongruous pairing of the artificial with the organic was part of the toys' fascination for journalists and newspaper readers; the chubby little ducks with their painted smiles navigating the brutal waters of the Arctic were, on the one hand, charming, and on the other hand, appalling. No matter how homey and familiar they appeared, no matter how comically juxtaposed the image of a plastic duck in an endless expanse of water seemed, no matter how amusing the color photographs or intricate maps in the *Daily Mail*, the toys were, after all, just garbage.

This fact was not lost on the reporters who found in the ducks a plethora of stories that seemed perfectly suited to account in any number of ways for ocean currents, pollution, beachcombing, and even the durability of the plastic toys themselves. The mainstream media coverage was thus rather eccentric: stories drew attention to the toys' strangeness or cuteness along with the financial rewards offered for finding one of the ducks, even as they also discussed big-picture issues like the common phenomenon of container ships losing cargo in the ocean or the effect of plastic on the ocean's food chain. Indeed, it seems that the more that the images of battered plastic toys were circulated, the more they inspired a surfeit of stories that, like the iconic plastic ducks themselves, seemed both familiar and unexpected. Their ghoulish charm appeared intimately tied to the fact that they were nonbiodegradable lumps of plastic, shedding bisphenol A and pthalates as they traveled the world.

Similarly, their visibility suggested even more upsetting questions: How many other objects have been dumped into the sea and gone unreported because they lack a hook, a cutesy angle that disguises the grim conditions of the role of global trade in planetary degradation? What about all the other unpublicized substances that get dumped, either by accident or on purpose, into waterways all the time? And where do all the millions of plastic toys that we buy every day end up when we throw them away? The rubber ducks' embodiment of polar opposites—at once commodities and trash, toys and junk, nature and artifice—made them interesting to reporters, but it also made them a sign of the role of garbage in embodying large- and small-scale problems of environmental crisis. As Curt Ebbesmeyer, a retired oceanographer who charted the ducks for over a decade, told the media with some ironic understatement, "You can learn a lot from a duck on a beach" (quoted in Clerkin 2007)[1]

The uncanny nature of these toys—the coordination of their domestic charm with their grim embodiment of global traffic and pollution—almost perfectly encapsulates what interests us most about garbage and waste. A deep concern with garbage, we discovered as we researched and organized this collection, is among the few environmental issues that seems to unite people who otherwise share very different vocabularies to express their environmental commitments. Studying garbage is as important to academics as it is to activists, as productive to the scholarship of humanists as it is to scientists, as critical for how social actors understand the habits of daily life as it is for how they grasp the economics of a global order of production, circulation, and consumption. Garbage and waste provide a way to begin to create large-scale models for understanding systems of consumption and production (What about the garbage and toxic waste we cannot see? How does nuclear waste affect the soil or water? How can it affect people far away?) as well as provoking a more finely grained understanding of the ecology of a town, neighborhood, or even individual household. The mark of what is despised after having been desired and needed, garbage and waste bring into sharp focus the complicated relationships between nature and culture. For this reason, garbage and waste occupy a unique position in how we narrate the keywords that underpin the cultural, social, and political effects and causes of an environmental crisis that brings global and local concerns into a sometimes-discordant conversation.

Indeed, as scholars like John Scanlan (2005) and Gay Hawkins (2006) have argued, notions of garbage and the terms associated with it are so deeply woven into the basic fabric of how culture operates they have come to be among the most powerful epistemological ways that human actors organize their sense of self along with their position in relation to various places, values, and other kinds of people. Because it is so ubiquitous, garbage is among the most immediate categories against which people are defined, and against which their identities as raced, gendered, and classed subjects are conceptualized. Yet as the authors in this volume discuss, although garbage gives many scholars and activists a shared starting point to analyze the relationship between the material of waste and metaphors that attend it, garbage is notoriously difficult to identify.

The *idea* of garbage—apart, say, from an inventory of the objects that a given individual evaluates, rejects, and throws away—is difficult to fully explain, yet therein lies its analytic utility, its capacity to bring conceptual oppositions into the same horizon of intelligibility. Garbage has a powerful ability to both materially embody and metaphorically indicate some of the most complicated ideas in contemporary studies of environmental politics and human cultures. Thus, our interest in garbage is focused on local exchanges between its material incarnations and symbolic iterations, each of which we understand to always work together to mediate different layers of social, political, economic, and individual experience. Garbage shapes subjective responses to both the things we throw out and the things we acquire, organizes how places and spaces are mapped, and points to contradictions in the way we define what counts as useful, valuable, wasteful, and unproductive as well as how we believe that those definitions naturally conform to given ideologies and discursive structures. Social actors classify objects, places, ways of thinking, kinds of discourses, and even other people as garbage, and those classifications, as sociologist Pierre Bourdieu reminds us, also classify them (1984). What does it mean, then, that garbage enables narratives of purity and contamination in the social and material worlds? How are such narratives made? Who makes them, and who suffers because of them? What do such narratives enable politically and socially? What kinds of subjects are produced in the crucible of narratives about waste?

For us, the central insight that garbage both bears within it contradictory stories and histories as well as embodies how people tell themselves stories about what kind of matter really matters inspired us to want to understand the interaction of the different narrative forms of waste and garbage. We are particularly interested in how different scholars define the stakes of how garbage is narrated and how it narrates other problems, and how they invoke keywords that enable others to participate in those narratives. Some of those keywords—value, utility, waste, surplus, pollution, contagion, disgust, damage, toxicity, circulation, destruction, and contamination—become apparent when we look closely at how objects (like the rubber ducks that were traveling the globe for over ten years) are central to various narratives. We hope that the contributors' emphasis on different incarnations of garbage will allow readers to see how specific material and metaphoric claims about what counts as garbage have real repercussions in a variety of contexts, locations, and scales.

The keywords we list above demand a certain analytic intimacy. How, though, is that intimacy achieved when we are discussing something that commonsensically so clearly inspires disgust, an immediate and nonanalytic reaction that seeks distance and boundaries? We suggest beginning with how that disgust itself produces the link between subjective responses to environmental offenses and a more critical investigation of them. Individual objects like the rubber ducks, to return to our example, not only prompt white papers on how to prevent ocean pollution, regulate global commerce, or deal with the endless supply of plastic toys from China. They produce subjective and affective responses—indignation, anger, anxiety, amusement, and disbelief—and these responses reveal the tension between individual relationships with the materiality of garbage, the identification of garbage as a category, and the corporate malfeasance that appears to be its primary source.

Garbage, especially if it is defined in the most basic and apolitical way as an object that has been deemed useless and cast aside, elicits mixed feelings from all kinds of people—feelings ranging from enchantment to outrage, often at the same time and about the same item. It is not surprising that such objects produce mixed responses, for those things that are broken or out of place have been, when whole, the repositories of feelings as well. We have wanted those things, loved them, needed them, saved for them (or bought them on credit and are still paying for them), and prized them, and now they have been used up, so we have to find something to do with them. When we pay attention to them, those discarded objects frequently ask us to reorganize how we respond to commodities and goods across space and time. And that, in turn, asks us to see that our public and private identities as citizens, consumers, and organic inhabitants of the globe are enmeshed with our responsibility toward the objects we discard, and furthermore, toward the systems that make those objects that we love and discard, and toward the subjects made by systems that make those objects, systems that abject some people and privilege others. The commodities that social actors in the rich world use and discard suture us into larger structures of families, neighborhoods, classes, and nations. But they produce actors' relationships to objects as not just social but also temporal. The commodities we fantasize about become the objects we can't figure out what to do with and that we hope will disappear, so that we can endlessly chase the new, pure, and more satisfying commodity.

The studies of garbage and waste that we present here richly illustrate how complicated the relationship between material and symbolic meanings can be, and exploring that relationship is the focus of our book. While none of the chapters in this volume are about rubber duckies, each addresses some aspect of the constellation of issues and keywords that those battered toys embody about the histories of what counts as waste and garbage, and about how those histories themselves become foundational for other ways of narrating waste, loss, and garbage in other social locations. Of particular interest to us as well as our contributors are relations between individuals and larger systems, and negotiations between the lived effects of waste and the stories told about those effects. Taken together, these chapters are especially interested in an account of how, as Zygmunt Bauman (2004) has recently argued, the individual human body and body politic can be coordinated around metaphors of waste and garbage; all bodies in late capitalism, according to Bauman, are out of place, and the political and aesthetic project of improving the self has made the relationship between waste and human value the central issue of our time. Although our contributors deal with a range of places and events, this volume as a whole consistently attempts to account for the seeming impossibility of reconciling individual affective responses to garbage with corporate responsibility to the environment. The book also offers some suggestions for relating the material aspects of garbage to its symbolic work.

Histories of the Dustheap was in part inspired by our fascination with the transformation of environmental issues into events about which social actors can—and in some sense *must*—have feelings. Doesn't the surfeit of anxiety about consumer choices, creeping self-righteousness about going green, and paralyzing despair about the impossibility of making a difference in the big scheme of things contribute to the wider abdication of political agency in an increasingly globalized world? As Michael Maniates (2002) and Thomas Princen (2002) have contended, the extraordinary emotional investment that environmental issues seem to elicit in people might offer an opening for corporations to make more money by appealing to the very ecosensibilities that the corporations themselves are helping consumers to develop, but those same emotional investments also seem to us to be a wasted opportunity to redirect those investments more effectively.

Converting the prosaic qualities of garbage into sustained and disciplined contributions can, we think, illuminate the overdetermined quality of the everyday experiences of garbage. Garbage is after all quite ordinary, produced daily by everyone as a by-product of merely being alive, but it is also extraordinary in its sheer volume and the practical problems it presents. For some people it is present and visible; for others it is a momentary irritation before it is hauled away. Sometimes garbage can be reclaimed, and made into something new and valuable—in which case, it disappears from the category of garbage and returns as a new commodity. Garbage is not just unused raw material, nor is it merely what something becomes when it is used up, nor does it only describe something once whole that has decayed or been broken. Garbage might best be understood not as an object but rather as a category, and it comes into being only at the moment when it is thrown away, when no other apparent use can be found for it. And yet no matter how many ways we experiment with how to define it—and virtually every book on garbage begins with a meditation on how to construct a definition—we are less interested in such definitions than in how and why they are deployed by different people along with how those definitions are pressed into the service of larger projects, ideas, and disciplinary commitments. The construction of massive public works projects to handle increasing volumes of sewage as well as residential and industrial garbage, for example, underwrites the very idea of the public as a material space of civic interaction—a space that ironically seems to have been overwritten by a belief in environmental commitments as private and personal.

The contrasts between the public and personal, the scholarly and everyday, in turn make visible the ways that the popular shift toward environmental thinking and action has sometimes obscured the larger social and industrial systems that contribute to the degradation of conditions in natural and built environments. These contrasts also highlight the fact that garbage is always most immediately apparent as a problem stemming from the most local forms of consumption—the things we see, buy, and throw away—that we can deal with on an equally local level. We can make garbage disappear materially and even metaphorically when we recycle, repurpose, transform a broken object into a new one, or even when we burn it as biomass. But what if transforming garbage doesn't really deal with the problems of waste?

Our authors together argue that garbage and the related issue of waste challenge conventional modes of seeing and understanding the world along with the objects we make within it. In much the same way as the image of the bleached and battered rubber duckies navigating the world's oceans interrupts received notions of what is good or bad for the environment (and what is useful and aesthetic, and the ethical webs those notions are caught in), we hope this volume invites the reader to analyze garbage in a way that interrupts the wasteful cycles of frustration about the incommensurability between the individual and corporate, the material and symbolic. As we describe, the contributors to this volume are interested in a myriad of definitions and varieties of waste, from wasted time, energy, money, and effort, to the waste products we commonly call junk, garbage, refuse, trash, rubbish, detritus, scrap, and shit. The relationship between waste and garbage, what we throw away and the stories we tell about what we discard, form the basis of this volume's substantive and methodological contribution.

We are, finally, interested in the stories about waste that a culture tells itself, for stories about what a culture throws away can explain what it values and disdains, what it believes about itself and how it relates to other cultures, and how it defines what kinds of people are valuable as well as what kinds of places are worthless. Stories have material effects, and the substance of lived reality prompts a variety of narratives. Encounters with garbage help demonstrate who has what kind of agency, and what range of possible actions people have available to them as citizens, consumers, and ethical inhabitants of communities in which the turn toward more sustainable practices have revealed systems and individuals to be far more interdependent than once thought. What all the chapters here have in common is the refusal to take a simple moral stance on the production of waste—all humans, after all, produce waste every day—and a commitment to waste's status as more than merely a metaphor in its relentless materiality. With this volume, we hope to interrupt frustrations and impasses regarding waste—not to necessarily correct them, but instead to change the terms of engagement and thereby make the conversations more productive.

II

Many books that came before ours deal with these and similar issues from various angles and theoretical perspectives, and many begin with a question of how to define garbage and waste theoretically, substantively and functionally. The keywords that we mention here structure how other scholars have discussed and defined the study of garbage. The chapters in this book have been informed by that work. Garbage and waste have a long-standing place in environmentally motivated scholarship, and choices about how to define garbage have been critical to the story each scholar tells about its political and social importance.

Scholarship on garbage—and scholarship that is organized around the study of garbage and waste—has concerned itself with the role of garbage and trash in the construction of large, infrastructural works, as a symptom of profound racial and gendered injustices, as the anthropological and theoretical basis of the subject's sense of self, and with how garbage organizes capitalism's drive toward creating a world in which commodities and consumption become the means of political expression. Garbage has been used to tell many different kinds of stories; it has been both metaphor and measurement, historical treasure house and collection of capitalism's failures. The ongoing scholarship on garbage—including the emergence of waste studies and public art projects on waste and garbage—in various environmental studies programs highlights how fascinating this most disgusting topic has been for researchers. We might begin a survey of how garbage and trash were seen to be constitutive by considering the impact of *Purity and Danger: An Analysis of Concepts of Pollution and Taboo* by Mary Douglas (1966). Douglas drew from the field of anthropology to make a powerful argument about social abjection and pollution by maintaining that women have functioned as disordering agents in culture—agents for whom being "out of place," as Douglas defined dirt, created the possibility of wider social disorder (p. 161). The fields of anthropology and literary criticism, including scholars who took up the challenge of postcolonial critique, built on claims about social disorder and its contaminating agents, and thereby helped to make visible the broad economic and cultural effects of long-standing and formative metaphors of garbage and refuse.

We might trace the emergence of studies that focus on the material nature of garbage in order to explicate larger theoretical models of subjectivity and the distribution of value according to defined cultural hierarchies to *Rubbish Theory: The Creation and Destruction of Value* by Michael Thompson (1979). Although out of print, Thompson's study remains in demand among scholars in the humanities and social sciences (if the recall wars at multiple university libraries are any indication). His work, like other studies that followed, consists of both theoretical discussions and case studies, and bases its argument on a definition of garbage that is usually determined by economic forces well beyond anything that could be called an object's inherent worth. Thompson's decision to pair a theoretical inquiry into garbage with case studies about its material incarnations has influenced many scholars and activists, who by now have reached some consensus that waste and garbage are both irreducibly material and yet culturally constructed and economically determined.

Paying attention to theoretical models and case studies has had the effect of making garbage an ironically productive category; if the objects of which garbage is composed are the broken or out-of-place commodities that seem to have fulfilled their intended function, historians and environmentalists have been able to "repurpose" it in order to tell stories about shifts in historical modes of creating, using, and transporting it. As scholars like Martin Melosi, William Rathje, and Cullen Murphy have shown, the increasingly sophisticated methods of waste disposal have had enormous effects on how our culture has come to define itself in ways that seem to have nothing at all to do with waste and garbage. In *Rubbish*, Rathje and Murphy (1992, 12) describe the work of the University of Arizona's Garbage Project, writing, "Human beings have over the centuries left many accounts describing their lives and civilizations. Many of these are little more than self-aggrandizing advertisements. . . . Historians are understandably drawn to written evidence of this kind, but garbage has often served as a kind of tattle-tale, setting the record straight." If garbage is a tattletale, countering the official record, it also gives us a way to unearth the histories of social actors who have often been left out of the "official history" in part because they were most affected by garbage. As Melosi (1981) argues in the introduction to *Garbage in the Cities*, we must see garbage and waste as a story about those people most affected by it, and must learn to narrate it in a way that uses their perceptions of it.

Similarly, historian Susan Strasser (2000) points out in *Waste and Want* that women have become the arbiters of what counts as filth and garbage because they have become the custodians of their homes' cleanliness and order. Here too, the historical story of waste gains much of its energy from the metaphoric resonance of waste, and even more so from its metonymic traces. Women's domestic labor makes them the arbiters of cleanliness and sanitation because, as cultural anthropologists (like Douglas) argue, women's bodies in particular are the symbolic boundary between what counts as purity and filth. But women, as Strasser also asserts, became empowered as the arbiters of what counted as valuable, was worth purchasing, should be consumed, and could be transformed. Their social status, in other words, put them in the position of making day-to-day judgments that had a powerful impact on how households and families developed historically, even as they became the prime targets of aggressive corporate campaigns that sought to persuade them to buy specific items and objects as an expression of "caring" for their families. If garbage can help narrate a more complicated history of gender in the United States, it also can provide a narrative about race and class. Sociologist David Pellow (2007) has contended that the sanitation and recycling industries have disproportionately affected working-class and ethnic minorities, and scholars who study the political ramifications of dump locations and toxic landfills have made similar claims. The interconnections between the metaphoric and historical work of waste and garbage thus also illuminate contemporary discussions of how garbage circulates and is sited in areas inhabited by politically and economically powerless people—people whose very association with garbage and waste makes them appear to be the source, rather than the victims, of contagion and contamination.

What fascinated us about the scholarship on garbage and waste was not merely the richness of its scholarly archive, or the liveliness of the debate about the stories that garbage tells. We became especially interested in how activists and scholars were trying to influence one another. We were intrigued to find that the use of garbage has become popular in environmental writing. For example, the recent *Garbageland: The Secret Trail of Trash* by Elizabeth Royte (2005) and *Gone Tomorrow: The Hidden Life of Garbage* by Heather Rogers (2005) were both widely reviewed. The authors were interviewed on National Public Radio and embarked on speaking tours. The books sold well too, in part because they were an attempt to make visible not only the work but also the life cycle of garbage. Each author tracked their domestic waste from garbage can to landfill, and found that waste hauling and landfilling are highly integrated industries, implicated in national, regional, and city politics and economies. Royte and Rogers also discovered that the handling of solid waste produces not only a range of jobs but relies on a range of technologies for landfill construction and maintenance as well. Along with the knowledge it yields, garbage also produces jobs and generates income. Indeed, the story of garbage is also the story of civilization writ small, for the same technologies that Royte and Rogers and investigate as they follow the route of household trash to its destination at a landfill have evolved from other models of waste disposal, as they detail in their books.

The popularity of their books was not a fluke—in fact, it testifies to the hunger that social actors feel for real information in the midst of so much corporate greenwashing—and we discovered that garbage and waste form the subjects of other popular nonfiction. The account of cholera in mid-nineteenth-century Britain in *The Ghost Map: The Story of London's Most Terrifying Epidemic,* by Steven Berlin Johnson (2007), for instance, details how efforts to define and clean up waste have been central to the rise of urban infrastructures as well as the emergence of public health and epidemiology. Similarly, recently published popular nonfiction about the conditions that create garbage in the first place aim to reduce garbage, pointing out that garbage from the West affects the health of populations in the emerging world. *Cradle to Cradle: Remaking the Way We Make Things* by William McDonough and Michael Braungart (2002), *Made to Break: Technology and Obsolescence in America* by Giles Slade (2006), and *High Tech Trash: Digital Devices, Hidden Toxics, and Human Health* by Elizabeth Grossman (2007) are all marketed to the same audiences that have read Royte and Rogers. Importantly, all of them expand the study of garbage to encompass globalization. Still other scholars who are more directly working in environmental studies have used the broader discourse of waste as a way to critique a capitalist system in which the global traffic in commodities makes those commodities barely separable from garbage, from the moment they are created in polluting factories, transported to far-flung places, and laid to rest in dumps all over the world. Princen (2002), Andrew Szasz (2007), and Maniates (2002) have diagnosed the problem of consumption—a problem that necessarily produces waste as a condition of its growth—as primary to how activist, popular, and scholarly commitments can be put into productive conversation.

Although we are heartened to see that the study of garbage has become a more popular topic—there are even YouTube documentaries about the dangers of e-waste for the Chinese workers who are usually tasked with recovering it—we believe that there is more work to be done to recover the histories of garbage and waste for a broad academic audience. In part, we think that the intersection between academic and popular writing has vitalized the study of garbage; Scanlan (2005), Hawkins and Stephen Muecke (2003; Hawkins 2006), and Dominique Laporte (2002) all work with complex poststructuralist theory, but in doing so, they take seriously the experience of everyday life, the way that social actors interact with the metaphors and material that create their worlds, and from which they create their sense of selfhood. Although each of these texts works with a highly sophisticated set of abstractions, they are eminently readable, provocative, and insightful contributions to the field. They address the ambivalent politics of recycling and repurposing as well as the affective attachment people feel about trying to make a difference in the sheer volume of trash. Like the scholars who come before us, we understand garbage as productive of subjectivities and constitutive of social relations, but our book aims to engage specific understandings of waste and garbage, rather than producing microengaged readings of various Foucauldian or Marxist models. Therefore, readers of this volume will find little explication of primary theoretical texts. Those familiar with poststructuralist theory will see it at work in many of the contributions. Those unfamiliar with theory will find engaged narrative and analyses concerning a range of garbage's effects.

We aim to chart similarities in the way scholars and activists talk about garbage, and promote conversations that can happen across as well as within disciplines. *Histories of the Dustheap* aspires to show that there is not only common ground but also multiple political and epistemological—as opposed to just disciplinary—ways of considering and narrating human problems. The multivocality of differently vested interests can sometimes be suppressed in academic discourse or obscured in much environmental reporting, or can be seen to be a problem to overcome rather than an inevitable by-product of human activity. For our part, we hope the collection provides a variety of voices and ideas. Thus, besides exploring relations between the material and symbolic actions of garbage, the book also implicitly examines relations between academic disciplines.

The dustheap of history (a phrase often attributed to Karl Marx, but actually coined by Leon Trotsky during some in-party disagreements, and recycled later by Ronald Reagan) should not necessarily refer to a place where things go to be forgotten, we think. Instead, our collection is premised on the idea that the dustheap of history is where we must look to confront our possibilities and failures, and where we must seek interruptions to entrenched and unproductive disavowals of how the material and symbolic work together when it comes to garbage. *Histories of the Dustheap* thus resuscitates the dustheap of history as a vital, productive site of human activity.

III

This collection is interested in finding a way to produce a common conversation among experts in humanistic and social scientific academic fields, coming from a range of methodological and theoretical backgrounds. As garbage embodies a story about the larger refusals and values of contemporary US culture, so too do the different approaches scholars take when discussing it provide us with critical information. How various scholars tell the story of waste is as important as the story that they believe waste itself tells, and this book is organized to bring those narratives into a single frame. Within the overall frame of exploring material and symbolic relations, each of the chapters in *Histories of the Dustheap* uses garbage, waste, and refuse to inquire into the relationships between various systems: the local and global, the economic and ecological, and the historical and contemporary, to give three examples. Garbage, waste, and refuse are not merely the most universal of materials. They have a special utility, for the category of garbage brings together amateurs and specialists as well as researchers and activists alike into a conversation about how the most prosaic, ugly objects can signal larger, more complex relationships between people and places by embodying human practices of consumption, production, and refusal. The book draws on the expertise of scholars, activists, and humanists with the aim of modeling the potential for experts from widely divergent fields to present information to one another and nonspecialist audiences clearly and cogently. Our book is founded on the premise that if the challenge of environmental action is to organize information in order to make it relevant for ethical action, the challenge of environmental scholarship is to present what has been specialized, disciplinarily defined information to a broad cross-section of interested, passionate readers.

In order to highlight our concentration on the symbolic and material aspects of garbage as it occurs in various disciplines and political projects, we divided the book into three sections, with each section devoted to one outcome of those negotiations: subjectivity, place, and ideological contradiction, respectively. Material and symbolic processes can be so interconnected as to render them often indistinguishable from one another, and indeed we only distinguish them here in the introduction to call attention to the book's treatment of them as constitutive of each other.

Thus, our first section, "The Subjectivities of Garbage," opens with Richard Newman's "Darker Shades of Green: Love Canal, Toxic Autobiography, and American Environmental Writing," which focuses on how the residents of Love Canal in Niagara Falls, New York, attempted to produce a counternarrative of their experience in the form of a toxic autobiography—a phrase that both renders visible the relationship of a self to an environment that shapes it and calls into question the purity of the memoir genre itself. Using the Love Canal case, Newman shows that although the narratives he studies are usually slow-burn tragedies, their emergence as a particular subgenre demonstrates how individual voices historically marginalized from large-scale environmental decision making can collectively bring attention to environmental catastrophes that are often ignored by government agencies. If this first contribution is about the fantasy of agency created through narrative, the next chapter looks at political agency created through a change in what we might call domestic infrastructure. William Gleason's "'The Most Radical View of the Whole Subject': George E. Waring Jr., Domestic Waste, and Women's Rights" focuses on one man's work on public and private drains in the late nineteenth-century United States, but it also elaborates a broader ideological story of what it means to connect a household to an urban infrastructure. Gleason's study shows that the history of public health is also the history of women's political rights, environmental responsibilities, and role in the home and public health. In the last work in this section, "Enviroblogging: Clearing Green Space in a Virtual World," Stephanie Foote describes how online communities and bloggers who concentrate on environmental issues must eventually come to recognize the limits of lifestyle changes as well as the potential for environmental community offered by the Internet. In this chapter, Michel Foucault's technologies of the self as a project of self-perfection meet the technology of the Internet around environmental issues, and reveal both the limits and possibilities of individual action in the blogosphere.

Our second section, "The Places of Garbage," explores the symbolic and material work that mapping along with other place-creating and place-managing strategies accomplish. The idea of place, perhaps even more than that of subjectivity, offers a glimpse into how the insistently material aspects of waste and garbage are symbolically handled. Because the idea of place seems so self-evident, it is an especially powerful example, as cultural geographers like David Harvey have demonstrated, of how ideological concerns define how we locate ourselves, and how we actively define while seeming merely to assent to how the natural and built environments are understood. Underwritten by raced, classed, and gendered assumptions of how space should "naturally" be understood and used, place is the site of struggles over power, knowledge, and individual identity as well as the apparent source of them.

Scott Frickel's "Missing New Orleans: Tracking Knowledge and Ignorance through an Urban Hazardscape" shows how municipal authorities created geographic zones of ignorance regarding the intensified soil contamination due to flooding from Hurricane Katrina. Frickel is concerned with how regulatory agencies create, sanction, and disseminate knowledge, and how that knowledge is used to address and suppress public concerns about contamination. The second contribution in this section shows the extent to which midwestern geography matters to how people deal with waste in Bloomington, Indiana. Phaedra Pezzullo's "What Gets Buried in a Small Town: Toxic E-Waste and Democratic Frictions in the Crossroads of the United States" picks up on some of the grassroots activism covered in the first section of the book. Pezzullo argues that grassroots activism cannot be based merely on an attempt to stop "bad technologies" but rather must rely on citizens rooted in their community who are willing and able to educate themselves about how corporate histories in that community have enabled as well as curtailed the health and welfare of its workers and citizens. Next, Elizabeth Mazzolini's chapter demonstrates the interplay between the cultural status of the well-known and highly culturally invested site of Mount Everest and the unsettled questions about what kind of problem garbage on Mount Everest might be, and how it changes the place of Mount Everest, both in the cultural imagination and on the ground. "The Garbage Question on Top of the World" shows how ambiguous knowledge can actually be beneficial to seemingly contradictory but similarly invested interests, as profitable sites get mapped and remapped according to cultural values.

Our third section, "The Cultural Contradictions of Garbage," highlights the kind of practical, theoretical, and political conundrums raised by garbage and waste. This last section pays particular attention to the moments when objects or commodities are in a state of transition, hovering between the state of becoming waste and remaining a valuable commodity. Each of the chapters here examines in some detail specific kinds of objects—sewage, toys, or plastic bags—along with the theoretical and practical debates that help to convert them from desirable commodities into trash. We thus include contributions that move from the specific to the abstract in order to demonstrate how different historical moments shape and are shaped by the relationship between what counts as valuable and useless.

In "Purification or Profit: Milwaukee and the Contradictions of Sludge," Daniel Schneider concentrates on that city's successes and failures over the course of the twentieth century as it attempted to purify its wastewater and sell its sludge as fertilizer. Schneider explores how industrial waste has been simultaneously a burden and resource for this one municipality, and demonstrates that the confluence of new agricultural techniques, civic improvement, and emerging business models made sewage, if only for a moment, among Milwaukee's most valuable products. Jennifer Clapp's chapter about the plastics industry shows the elaborate inconsistencies that the plastics industry endorses in order to protect its business interests. "The Rising Tide against Plastic Waste: Unpacking Industry Attempts to Influence the Debate" shifts attention to global corporate responses to widespread consumer tastes and practices. The permanence and ubiquity of plastic bags and bottles makes them an environmental scourge, the primary symptom of which are the now well-known masses of plastic found in the world's oceans. Clapp describes how the plastics industry works to protect its interests by maintaining a counterintuitive discourse about plastics rooted in cultural values like "freedom," "convenience," and "choice." Lastly, Marisol Cortez engages with issues of consumer desire and revulsion to show the complicated relationship between nostalgia and neglect. "Time Out of Mind: The Animation of Obsolescence in *The Brave Little Toaster*" is a fascinating look at an obsolete movie about obsolescence. Set within a larger critique of the entertainment industry's relentless drive to create the next new thing, Cortez argues that the film's plot of imagining the secret life of everyday things reveals the commodity fetishism that obscures the actual secret life—or material relations—of these same kinds of objects in the world of the viewer. Although Cortez recognizes that affective attachment to waste does not immediately solve the injustices of unequal material and political relations, her chapter begins a conversation about what it means to keep objects and invest them with meaning that is not limited to their mere economic value, and that embraces their contradictory places in our lives.

Note

1. For two particularly emblematic stories of the plastic animals, see Clerkin 2007; Ford 2003. The *Daily Mail* and *Christian Science Monitor* have rather

different journalistic missions, but their reporters, Clerkin and Ford, each pick up on the uncanny nature of the ducks in their respective stories. For a full account of the incident, see Hohn 2011.

References

Bauman, Zygmunt. 2004. *Wasted Lives: Modernity and Its Outcasts*. Malden, MA: Polity Press.

Bourdieu, Pierre. 1984. *Distinction: A Social Critique of the Judgment of Taste*. Cambridge, MA: Harvard University Press.

Clerkin, Ben. 2007. Thousands of Rubber Ducks to Land on British Shores after 15-Year Journey. *Daily Mail*, June 27. http://www.dailymail.co.uk/Thousands-rubber-ducks-land-British-shores-15-year-journey.html (accessed June 10, 2011).

Douglas, Mary. 1966. *Purity and Danger: An Analysis of the Concepts of Pollution and Taboo*. New York: Routledge and Keegan Paul.

Ford, Peter. 2003. Drifting Rubber Duckies Chart Oceans of Plastic. *Christian Science Monitor*, July 31. http://www.csmonitor.com/2003/0731/p01s04-woeu.html (accessed June 10, 2011).

Grossman, Elizabeth. 2007. *High Tech Trash: Digital Devices, Hidden Toxics, and Human Health*. Washington, DC: Island Press.

Hawkins, Gay. 2006. *The Ethics of Waste: How We Relate to Rubbish*. Lanham, MD: Rowman and Littlefield.

Hawkins, Gay, and Stephen Muecke, eds. 2003. *Culture and Waste: The Creation and Destruction of Value*. Lanham, MD: Rowman and Littlefield.

Hohn, Donovan. 2011. *Moby-Duck: The True Story of 28,800 Bath Toys Lost at Sea, and of the Beachcombers, Oceanographers, Environmentalists, and Fools, including the Author, Who Went in Search of Them*. New York: Viking Press.

Johnson, Steven Berlin. 2007. *The Ghost Map: The Story of London's Most Terrifying Epidemic*. New York: Riverhead Trade.

LaPorte, Dominique. 2002. *History of Shit*. Cambridge, MA: MIT Press.

Maniates, Michael. 2002. Search of Consumptive Resistance: The Voluntary Simplicity Movement. In *Confronting Consumption*, ed. Thomas Princen, Michael Maniates, and Ken Conca, 199–236. Cambridge, MA: MIT Press.

McDonough, William, and Michael Braungart. 2002. *Cradle to Cradle: Remaking the Way We Make Things*. New York: Northpoint Press.

Melosi, Martin V. 1981. *Garbage in the Cities: Refuse, Reform, and the Environment, 1880–1980*. College Station: Texas A&M University Press.

Pellow, David. 2007. *Resisting Global Toxics: Transnational Movements for Environmental Justice*. Cambridge, MA: MIT Press.

Princen, Thomas. 2002. Consumption and Its Externalities: Where Economy Meets Ecology. In *Confronting Consumption*, ed. Thomas Princen, Michael Maniates, and Ken Conca, 23–42. Cambridge, MA: MIT Press.

Rathje, William, and Cullen Murphy. 1992. *Rubbish! The Archaeology of Garbage*. New York: HarperCollins Publishers.

Rogers, Heather. 2005. *Gone Tomorrow: The Hidden Life of Garbage*. New York: New Press.

Royte, Elizabeth. 2005. *Garbageland: The Secret Trail of Trash*. New York: Little, Brown and Company.

Scanlan, John. 2005. *On Garbage*. London: Reaktion Books.

Slade, Giles. 2006. *Made to Break: Technology and Obsolescence in America*. Cambridge, MA: Harvard University Press.

Strasser, Susan. 2000. *Waste and Want: A Social History of Trash*. New York: Holt Paperbacks.

Szasz, Andrew. 2007. *Shopping Our Way to Safety: How We Changed from Protecting the Environment to Protecting Ourselves*. Minneapolis: University of Minnesota Press.

Thompson, Michael. 1979. *Rubbish Theory: The Creation and Destruction of Value*. New York: Oxford University Press.

I

The Subjectivities of Garbage

1

Darker Shades of Green: Love Canal, Toxic Autobiography, and American Environmental Writing

Richard Newman

Like a recurring nightmare, Robinson Kelly vividly recalls the toxic morning of July 16, 1979, when waves of polluted water poured down the Puerco River near Church Rock, New Mexico. "I didn't know what was going on but it was an ugly feeling," he told the *Navajo Times* at a rally commemorating the event. "I went to work and found out the dam broke." The cracked dam, owned by United Nuclear Corporation, spewed ninety-four million gallons of mining refuse, including eleven hundred tons of uranium waste, along eighty miles of river way. It was "the single largest release of radioactive material in US history," according to one estimate—far worse than Three Mile Island. Decades later, Kelly hoped his story would remind Americans that nuclear waste issues still plagued Indian country, endangering not only the land but also the people who never left it.[1]

At roughly the same time that hazardous waters flooded Kelly's home, a woman thousands of miles away described a similar toxic experience. Anne Hillis, a young mother residing in the Love Canal neighborhood of Niagara Falls, New York, discovered that a dump containing over twenty thousand tons of caustic chemical residues sat just a few blocks from her house. "People of the area are on the verge of hysteria," she wrote in spring 1979, months after state and federal emergency declarations had evacuated only a fraction of the neighborhood's families. In a hard-hitting (though unpublished) memoir titled "Love Canal's Contamination: The Poisoning of an American Family," Hillis detailed her debilitating experiences inside the toxic zone—experiences that almost drove her insane. "When you live in hell," she declared, "you have no fear of death! It's the living that you fear."[2]

As Kelly's and Hillis's narratives both illustrate, a new genre of US environmental writing emerged in the final decades of the twentieth century: toxic autobiography. Less concerned with natural landscapes and wildlife preservation than previous generations of environmental works, toxic autobiography meditates on the personal, political, and historical meanings of the hazardous waste grid that developed in the United States following World War II. Inextricably linked to the cause of environmental justice, toxic autobiography has become a more prominent part of the media landscape in recent years. Yet despite slick Hollywood renditions of toxic tales (including movies like *Erin Brockovich* and *A Civil Action*), the story remains much the same for generations of toxic writers: they seek to uncover life on the environmental margins. Here, away from breathtaking views that dominate many nature preservation struggles (and hence, away from mainstream ideas of what constitutes environmentalism), toxic autobiographers offer disturbing portraits of incinerators that infect urban air, brownfields whose corroded infrastructure threatens nearby fields and streams, and hazardous waste dumps that litter struggling communities.

This chapter examines toxic autobiography's ascension as a distinct genre of environmental writing. Surveying some of the most significant work from the late 1970s to the present, I argue that toxic autobiography flows from a deep sense of crisis among marginalized groups of people—including working-class communities, ethnic and racial minorities, and women—that feel trapped in poisoned landscapes well beyond mainstream concern. Looking for ways to put hazardous waste issues on the nation's political radar, they have turned to one of the oldest and most effective tools in global reform: personal narratives of suffering and pain. As a literary form dedicated to the truth telling of toxic experiences, the genre has allowed people separated by time, background, region, class, and ethnicity to see themselves as part of a common environmental cause.

But to what end? As I seek to show, generations of toxic autobiography have impacted contemporary environmental politics in several ways. First, as a subset of ecocriticism, it has helped return environmentalism to its core values of literary outreach. Much like earlier generations of nature writers (who sought to preserve wild nature), toxic autobiographers have relied on storytelling to convey their outrage about environmental degradation within industrial culture and society. Second, toxic autobiography has contributed to national and global redefinitions of environmental reform as necessarily encompassing vernacular environments—the daily, and often polluted, landscapes where most people live, work, and raise families. For toxic writers (many of whom are women), the air, water, and land of our everyday environment must be protected no less than that of pristine nature beyond the human grid. As both a form of ecocriticism and a tool of grassroots environmentalism, toxic autobiography has made visible industrial modernity's hazardous dustheap—a place once thought to be empty, and simply waiting for chemical and nuclear detritus, but now known to contain masses of people struggling with environmental health consequences. Literary outreach became, for them, the cornerstone of a new and toxic environmental sensibility—one that has ultimately impacted state, national, and even international politics.

Foundations of Toxic Literature

How and when did toxic autobiography start? Like other vanguard genres, it flowed from a variety of influences. In one sense, toxic autobiography is a logical extension of what Lawrence Buell (2005, 22) has termed "second wave eco-criticism." Focusing on the significance of vernacular landscapes (from the urban core to military-industrial grids), second-wave ecocritics ascended in the 1960s and 1970s, when environmentalism itself was reborn. As novelists, essayists, and even activists, they viewed built environments as key sites of US environmental thought. While not properly described as ecocritics (few are academics or write at the intersection of literary criticism and environmentalism), toxic autobiographers would certainly agree that industrialized, urbanized, and polluted human landscapes provide important insights about US environmental values.[3]

No second-wave ecocritic, of course, had more influence on toxic autobiography than scientist and best-selling author Rachel Carson. Author of the landmark book *Silent Spring*, Carson argued that neighborhoods, parks, and roadways formed a critical part of the modern US environment—an environment under siege by a postwar petrochemical complex. In her book's famous opening chapter, "A Fable for Tomorrow," Carson imagined a town denuded of its songbirds by chemical despoliation. "No witchcraft, no enemy action had silenced the rebirth of new life in this stricken world," she wrote. "The people had done it themselves." Through an unquestioning reliance on chemicals, insisted Carson (1962, 14), human populations might soon seal their own fate too.[4]

Carson's dystopian vision inspired others to provide real-world analogs to *Silent Spring*. From *Our Poisoned Earth and Sky*, a book on chemicals in the food chain by wellness guru J. I. Rodale (1964), to *Ecocide . . . and Thoughts toward Survival*, a set of academic essays, edited by Clifton and Jean White Fadiman (1971), from a 1970 conference on the coming age of environmental ruin, a range of environmental reformers tried to alert

Americans to the unnatural disasters looming in their neighborhoods, backyards, and gardens. "How effective is the control of water pollution in your state?" Rodale (better known today as the founder of *Prevention Magazine*) challenged readers. "Is anything being done about the . . . smog emanating from cities and highways in your part of the country? . . . [W]hat can you do to protect yourself from mass spraying? Excessive medical radiation? . . . [H]ave we reached the point of no return?" (Rodale 1964, 9). *Ecocide*'s editor did not even ask such questions; the verdict had arrived. "The environment [is] being murdered by mankind," Fadiman (1971, 9) declared. "Each day brings to light a new ecological crisis. Our dense amber air is a noxious emphysema agent; farming . . . turns fertile soil into a poisoned wasteland; rivers are sewers, lakes cesspools, and our oceans are dying." Part of what the political scientist Charles O. Jones (1972, 590) labeled the "eco-scare" genre of the early 1970s, *Ecocide* aimed to compel Americans to embrace environmentalism as the last best hope of averting an "ecosystem depression" (Fadiman 1971, 9).

While these dire perspectives helped postwar environmentalism lobby effectively for a wave of new laws during the 1970s (most notably, the Clean Air and Clean Water acts), toxic waste issues did not receive similar attention until later. In fact, the Resource Conservation and Recovery Act of 1976, which regulated hazardous waste production from "cradle to grave," did not have any regulatory teeth until the 1980s. And other laws aimed at reducing and/or regulating hazardous waste were phased in slowly by federal officials. For instance, the 1977 amendments to the Clean Water Act gave industry nearly a decade to comply with stricter regulations of river pollution. Even the Superfund, the federal government's signature program for dealing with problematic toxic waste sites, would not be created until December 1980.[5]

Frustrated that toxic trouble did not rate earlier action, new groups of reporters, activists, and writers sought to put hazardous waste issues in the forefront of the public's mind during the late 1970s. Indeed, many early toxic waste activists believed that mainstream US environmentalism had become too professionalized and overly concerned with wildlife issues. At Love Canal in the late 1970s, a young mother named Lois Gibbs found that many so-called big enviros (national environmental organizations) did not view toxic waste as a winnable issue in state or federal politics.[6]

As a result, toxic autobiographers also looked beyond mainstream environmentalism for sources of inspiration and support. Social movement literature loomed large here, particularly the writing of labor activists, abolitionists and civil rights reformers, and first- and second-wave feminists. In each case, as in toxic autobiography, literary activism undergirded the struggle for justice, compelling US citizens to take note of damaging social practices in their midst. Whether in the nineteenth-century example of Harriet Beecher Stowe's *Uncle Tom's Cabin* or the twentieth-century one of Upton Sinclair's *The Jungle*, activist writing personalized injustice via vivid images and tales of woe, taking people to "slave row" or inside the factory of industrial horrors. Activist literature, for toxic autobiographers, provided usable models for their own struggle to illuminate lives that many people simply did not see—or did not want to see.

As a matter of fact, toxic autobiography first took shape as investigative journalism. No better example exists than the work of Michael H. Brown, a Niagara Falls native who first covered the Love Canal saga unfolding in his own backyard during the late 1970s. After revelations in 1978 that the Love Canal neighborhood had been built on and around a buried toxic waste dump, Brown met with terrified residents over the course of several months. He filed over a hundred stories on the crisis, becoming the archetypal crusading environmental journalist. His visceral reporting style—influenced by the so-called new journalism in which correspondents inserted themselves into the story they covered—brought toxic landscapes alive for audiences locally and nationally. Brown took readers inside people's homes, where chemical waste oozed down basement walls, and into their backyards, where corroded barrels pushed through the soil. Populated by anxious, angry residents who had experienced miscarriages, cancers, and a host of other troubling ailments that conjured nothing less than the biblical story of Job, Brown's toxic neighborhood embodied the modern environmental dystopia. Who could live here, he wondered?[7]

After Love Canal, Brown reported on toxic problems in Michigan, New Jersey, and Pennsylvania, illuminating a hazardous waste world that few people knew existed, and even fewer understood. But by the early 1980s, his journalistic portraits of "Love Canal, USA" were merely the vanguard of a growing roster of books examining the nation's toxic landscape. Writers in these works emphasized more than the technical, scientific, and/or philosophical issues involved in toxic waste. For them, the human impact often proved equally, if not more, important. Ominous titles like *Poisoned Land* by Irene Kiefer predicted that toxic "time bombs" were "waiting to explode" across the United States, and with them a never-ending public health nightmare. "The episode at Love Canal is one of the worst environmental disasters ever," Kiefer, a former chemical industry insider, observed. "But it is only part of the story." The deeper story lay in the thousands of uncontained hazardous waste sites across the United States. How would people living near old toxic dumps respond when told they occupied the next Love Canal? Though she struck a positive note about the possibility of regulating hazardous waste in the future, Kiefer (1981, 3–12) stated bluntly that "completely cleansing our poisoned land of its lethal burden will take many years and billions of dollars."

David Morell and Christopher Magorian, policy wonks who moved between academic and political worlds, were not so sanguine. Believing that toxic waste disposal had reached a crisis point by the early 1980s— with more communities rising in protest over the dumping of hazardous materials in their midst—they called for far more open placement policies. Failing that, they argued in *Siting Hazardous Waste Facilities*, the nation's toxic problem would grow worse, with more communities exploding like Love Canal.[8]

Although Brown, Kiefer, and Morell and Margorin stressed the way that hazardous waste had become a palpable public issue, their books still had an "outside in" perspective. They were speaking *about* local communities, not as members of the toxic body politic. In fact, only Brown's work detailed the innermost struggles of local people, especially the way their lives had been reshaped by toxic landscapes. But many matters remained unresolved. How did people themselves in these communities describe and respond to their toxic existence? What medical and psychological impacts did hazardous waste produce on those living in the toxic zone? And just where were the nation's toxic hot spots?

From Revelation to Activist Writing: The Early Rise of Toxic Autobiography

By the early 1980s, a series of hazardous waste revelations made these questions vital in communities stretching from Louisiana to Michigan and Massachusetts to New Mexico. Research by state and federal investigators revealed that well over thirty thousand uncontained dumps existed in the United States, over a thousand of which were potentially hazardous. As an ABC News documentary called *The Killing Ground* (1979) showed, many communities were now dealing with Love Canal–type concerns; it was a national, not local, problem.[9] The 1983 evacuation of over two thousand residents of Times Beach, Missouri (southwest of Saint Louis) in the wake of widespread dioxin exposures further fueled fears of a toxic cloud hanging over US homes.

By then, reporters and environmental experts were not the only figures interested in toxic places. Love Canal, Times Beach, and other unnatural disasters served as a narrative launching pad for a new generation of community stakeholders—grassroots activists who gave voice to their struggles inside poisoned places. The Citizens' Clearinghouse for Hazardous Waste (CCHW), a national grassroots organization formed in the wake of Love Canal, recognized the great potential of community narratives in the struggle for environmental justice. If local people could tell their stories, stronger hazardous waste policies might well be the result. The group's "Leadership Handbook on Hazardous Waste," a self-published pamphlet, encouraged community mobilization as the most effective means of putting toxic matters on local, state, and even national political agendas. While offering tips about investigating toxic dumping grounds, forming activist organizations, and dealing with both political officials and business representatives, the handbook also told community activists to think of themselves as the architects of a powerful grassroots environmental story. "Ultimately, through your own work, you will be writing your own book," the pamphlet cheered (CCHW 1983, 1–2; see also 55–60).

It is no coincidence that one of the CCHW's founders had already created a model toxic autobiography. Gibbs's 1982 narrative, *Love Canal: My Story*, became perhaps the most important first-person environmental narrative of the late twentieth century. Written in an accessible style that would appeal to people with little formal education, her tale illustrated how average citizens mired in hazardous circumstances could confront both political and corporate powers. Gibbs remained an inspiration for subsequent writers not simply because she authored the inaugural toxic autobiography but also because she framed her struggle in terms of grassroots authenticity. "Almost everyone has heard about Love Canal," she began her narrative, "but not many people know what it is all about." Tellingly, the original paperback edition featured a cover with an epigraph that evoked Gibbs's central purpose: "Love Canal: Ask Those Who Really Know." Using the line from a Love Canal Homeowners Association (LCHA) handout, Gibbs indicated that notwithstanding considerable media attention, Love Canal remained a story defined by residents' experiences. "The most powerful thing you have is your story," Gibbs (1982, 1–2) told others seeking to highlight the perils of living in the toxic zone.[10]

Gibbs also showed how she and her neighbors moved from concern about toxic surroundings to grassroots environmental activism.[11] As a young mother of two who had recently moved to the largely working-class subdivision of Love Canal—a neighborhood of nearly a thousand families, many who worked in nearby chemical and manufacturing plants—she initially loved the area. "If you drove down my street *before Love Canal*," Gibbs (1982, 8) wrote, "you might have thought it looked like a typical American small town that you would see in a movie—neat bungalows, many painted white, with neatly clipped hedges or freshly painted fences." She and her husband did not worry about environmentalism or toxic waste. Gibbs even discounted initial news stories about the Love Canal dump, thinking they could not possibly correspond to her home. Yet her son Michael's health concerns, combined with increasing coverage of Love Canal, compelled Gibbs to pay closer attention to her local landscape. In the summer of 1978, researching her neighborhood's past, she and other residents realized that the dump was enormous—containing over a hundred thousand barrels of waste—and potentially more toxic than anyone had known or admitted. In fact, her son attended a school that had been built literally on top of the covered-over dump.[12]

Punctuated by a brand of folk wisdom that would appeal to average US readers (she tells her brother-in-law scientist explaining the environmental impact of chemicals to "translate some of that jibber-jabber . . . into English"), Gibbs's narrative takes readers through her education as an activist. When she thought about "my son attending that school," Gibbs grew "alarmed." Still, if local officials had merely moved Michael to a new school, Gibbs might not have become an activist. But their intransigence—and her family's growing list of ailments—compelled Gibbs to organize other mothers and families into a group: the LCHA. Emphasizing again and again the ups and downs of grassroots activism (from the thrill of planning rallies and challenging political leaders to the disappointment of telling neighbors that they must remain in their toxic homes), Gibbs makes it clear that there was no visionary public policy at Love Canal. Only the gritty and grueling mobilization of residents (and their allies) succeeded in keeping Love Canal alive in the public mind through the years. By the time state and federal officials finally offered permanent evacuation to area residents in late 1980, Gibbs felt not so much relieved as reborn. Looking back, she declares that Love Canal stood for a new understanding of democracy and environmentalism—one in which corporate power does not overwhelm the "public's right . . . to a safe environment" (Gibbs 1998, 26–27, 12).

Gibbs was not alone in expressing a larger sense of environmental purpose. In oral histories, congressional testimonies, memoirs, and written correspondence, other Love Canal residents made similar claims. Between

1979 and 1980 alone, for instance, nearly a dozen Love Canal residents testified before various federal officials, depicting the myriad health concerns, psychological traumas, and emotional fears that haunted their lives and turned them into grassroots environmental reformers. Grace Mc-Coulf, a LCHA volunteer, came to see herself as part of an environmental vanguard on hazardous waste issues. "With an educated population pushing the politicians and bringing to light all the potential problems related to hazardous dumping," McCoulf (1979; emphasis in original) told Congress, we "can help [politicians] with new and more effective laws—laws which will be and MUST be enforced to their fullest!!!"

Luella Kenny, a researcher at Roswell Park Memorial Cancer Institute in nearby Buffalo, offered an equally powerful vision of the way her toxic landscape drove her to activism. Playing on the American dream that one's home should be a sanctuary, Kenny told congressional officials in May 1979 about the way chemicals had overtaken her life. Like other Love Canal residents testifying before federal politicians, Kenny's brief but moving story (which she repeated to chemical company executives the following year) focused on the formerly unknown toxic borders surrounding her house. Her big backyard near a bucolic creek had once been a playground for Kenny's happy family. Her son Jon enjoyed splashing through backyard puddles formed by periodic floods. Yet when Jon came down with an atypical form of the kidney disorder nephrosis, which eventually killed him in October 1978, Kenny realized that chemicals had invaded her home. She learned that hazardous waste from Love Canal had migrated through sewers that emptied into the flooding creek. As if to underscore the reality of a "Silent Spring" at her Love Canal home, Kenny offered a striking visual recollection of a bird that drank from the creek one day and then dropped dead right in front of her. If state health officials had not been present as well, she stated, her toxic vignette would have been merely fanciful. Now it seemed prophetic, symbolizing the toxic reality of her local environment. Her home, bounded by chemical threats, had become a terrifying landscape. Joining neighborhood activist groups, she began marching at rallies, speaking to politicians and health officials about toxic waste, and envisioning herself as a Tom Paine–style revolutionary raising consciousness about basic US rights—in this case, the right to a clean living environment (see Kenny 1979).

These and other stories helped convince other victims of hazardous waste to speak out. No sooner had Love Canal residents mobilized, testified, and constituted their struggles in print than grassroots activists around the country started contacting them for advice and support. In

East Gray, Maine, for instance, a woman named Cathy Hinds formed the Environmental Public Interest Coalition in 1979 after communicating with Gibbs about the chemical contamination of her local aquifers. How, Hinds asked, can I raise awareness about and pass laws regulating hazardous waste dumps in my own neighborhood? Form a group and tell your story, Gibbs responded. With that advice in mind, Hinds mobilized her neighbors and then testified before Maine legislators about the need for a law "monitoring . . . toxic chemicals from the beginning of manufacturing all the way through waste disposal." Her gripping testimony revolved not around technical expertise but rather personal experiences in the toxic zone—specifically the tragic death of her infant son (which she linked to benzene poisoning). In a bare-bones recollection of her toxic struggles, Hinds remembered how she "broke down crying" when testifying before local politicians about the memory of her son's death. All Hinds "wanted to do was get up and run out of the room"; but she persevered and "manage[d] to tell her story." The law was passed, and "someone told [Hinds] that no bill had ever been pushed through committee so quickly" in Maine.[13]

As these early toxic tales illustrate, gender figured prominently in the rise of environmental justice movements around the country. "Women are more likely than men to take on such issues" as the chemical pollution of neighborhoods and homes, environmental justice activist Cynthia Hamilton has pointed out. Though women certainly work outside the home in many working-class and/or ethnic communities, domestic spaces (homes, schools, and streets) have "often been defined and proscribed as women's domain." As wives, mothers, and community caregivers, women see the day-to-day impact of toxic pollution on neighborhood children and families. And they usually become the first to speak out about the perils of hazardous waste. Hamilton (2000, 542) herself spoke from experience as a housewife turned activist in South Central Los Angeles. During the 1980s, city officials planned to operate a hazardous waste incinerator in Hamilton's neighborhood. She and other women organized against its placement, arguing that the incinerator would tarnish the air quality by releasing hazardous particulate matter into the South Central environment. Not only did Hamilton's mobilization of neighborhood women succeed; her story also bucked tacit but powerful industry beliefs that marginalized communities—populated by people in lower lower-income and educational strata—could not, and therefore would not, stand firm against the siting of toxic dumps and incinerators.

Like women's efforts at Love Canal, Hamilton's activism redefined urban landscapes as a key part of nascent environmental justice struggles.

Just as important, her narrative added another layer to a growing body of written work from inside the toxic zone.

Toxic Autobiography and Geocriticism: Remapping the Toxic Landscape

If telling a toxic story allowed grassroots stakeholders to put both themselves and hazardous waste issues on the nation's environmental radar immediately following Love Canal, then the genre also helped activists reimagine the very nature of the toxic grid. Critiquing the idea that hazardous waste was simply out of sight and out of mind (or a minor matter when compared to other environmental issues like resource conservation or wildlife protection), toxic autobiographers and their allies began charting the political contours of hazardous landscapes—their formation, socioeconomic meaning, and poisonous imprint on specific communities. For many writers, in fact, geography defined one's toxic destiny. And it was no mistake that certain locales seemed destined to be hazardous waste receptors. "There is mounting evidence," Gibbs observed, that "communities are . . . chosen for [the] siting [of] a waste disposal or industrial plant based on their demographics." Citing reports by industrial consulting firms that marginalized groups lacked the "community power" to oppose the toxic facilities in their midst, Gibbs (1982, 1–2) noted that grassroots activists realized early on that the "environmental and public health threats" born of the toxic zone were not "random." Certain people and places were at greater risk than others.

In Native American writers' variations on the theme, US hazardous waste policy epitomized long-held beliefs that, as one recent literary critique summarizes it, "the lands of indigenous peoples are underdeveloped and empty" (quoted in Reed 2009, 28–32), and therefore acceptable as toxic dumping grounds. Recent studies have shown that perhaps 20 percent of uranium-mining sites (and the waste they produce) sit on native reservations and land trusts—a disproportionately high figure for a population that accounts for less than 1 percent of the nation's total.[14] As Native American activist and writer Winona LaDuke has asserted, this toxic conception of Native American geography continues to shape contemporary nuclear waste discussions. "What happened when the best scientific minds and policy analysts in the world spent twenty years examining every possible way to deal with the problem of nuclear waste?" LaDuke asks. They recommended that all nuclear waste be transferred "thousands of miles . . . and dump[ed] . . . on an Indian reservation" at

Yucca Mountain. Yet "Indian people refused to become the silent martyrs of the nuclear industry," LaDuke (Grinde and Johansen, 211) continues. "We stand fighting in our homelands for a future free of the threat of genocide for children."

As both a chronicler of "Native environmentalism" and a storyteller/ autobiographer, LaDuke engages in something I call "geocriticism." More than an attempt to chart toxic waste zones, geocriticism rejects environmental policies that divide the world into appropriate and inappropriate dumping grounds. LaDuke expresses outrage at the various ways that mainstream culture has linked marginalized populations (not just Native American groups, but people of color, working-class communities, and inhabitants of the farm belt as well) to the toxic zones they inhabit. In a very real way, she argues, the modern toxic grid has formed out of a belief that certain people can be sacrificed to industrial progress—that they are the environmental equivalent of the deserving poor.

For LaDuke and many toxic autobiographers, remapping the hazardous waste grid compels Americans to see the vernacular landscape with new eyes. Toxic waste does not magically disappear; it goes into dumps, incinerators, and transfer stations located in the heart of certain (often-poor) communities. In this manner, toxic waste has corroded societal values by making environmental risk acceptable—as long as down-and-out communities have to deal with it. At Love Canal, residents did more than lash out at this notion. They documented their toxic grid to illuminate the day-to-day risks associated with their environment. The LCHA created maps of residents' illnesses, miscarriages, and deaths; charted the flow of waste in area sewers and streams; and expanded definitions of hazardous zones outward from the old dump itself in an attempt to convince state and federal officials that the entire Love Canal neighborhood—not merely the homes bordering the old dump—constituted a disaster area.

Gibbs recalled how residents helped innovate "the swale theory"— which linked disease patterns to the chemically laden wet areas, or swales, on which homes had been built—via maps and diagrams. Tracking results from a state health survey that LCHA members worked on, Gibbs and the LCHA noticed a correlation between health problems and swales, probably old streambeds and wetlands where chemicals from the dump had migrated. "We took out our health survey notebook and started to put squares, triangles, and stars on a street map, with a different symbol for each disease," Gibbs wrote (1982, 66), and when they had transposed the swales onto the demographic grid, "suddenly a pattern emerged!"

That map was eventually published in the *Niagara Falls Gazette*, creating a palpable image of people mired in hazardous waste.[15]

Love Canal activists turned these maps and images into a gripping David-and-Goliath narrative, contrasting their tough socioeconomic circumstances with the seemingly apathetic actions of state and federal officials, who moved slowly against the perils of hazardous waste. Would similar delays have occurred in wealthier neighborhoods? Would a toxic dump even get placed there? By constantly emphasizing the "working-class" quality of the community, Gibbs portrayed Love Canal itself as the archetypical underdog seeking nothing more than justice and a fair shake from disinterested political officials. This depiction was not completely accurate—there were professionals in the Love Canal neighborhood, and even Gibbs referred to her community as "middle class" on occasion—but that mattered less than the fact that Love Canal activists put a name and face on the problem of hazardous waste disposal.

Grassroots activists in other parts of the country further explored the link between hazardous geography and marginalized status. In Houston, a group led by attorney Linda M. Bullard and sociologist Robert Bullard filed a class action lawsuit in 1979, alleging that city officials practiced environmental racism. As Robert Bullard later recalled, *Bean v. Southwestern Waste Management* became "the first . . . in the United States that charged environmental discrimination in waste facility siting under the Civil Rights Act." To bolster this claim, the Bullards conducted prodigious research at Houston's administrative archives and interviewed a variety of local residents about the location of dumps throughout the famously unzoned metropolis. They concluded that black Houstonians were targeted as outliers—inhabitants of the toxic zone. Routinely, Bullard (1990; xvi) would write in *Dumping in Dixie*, a book pitched not just to academic specialists but also citizens for whom such toxic issues still remained unknown, the residential character of the neighborhoods had been established "long before the industrial waste facilities invaded them." For the Bullards, toxic pollution itself became a form of geographic segregation.

The Bullards' legal and academic work prefigured a 1987 report by the United Church of Christ on the link between racial and toxic geographies. The result of five years of work, the report found that "racial and ethnic Americans are far more likely to be unknowing victims of exposure to . . . [toxic] substances." In Atlanta, for example, the study pointed out that the majority of "uncontrolled toxic waste sites" existed in areas where the black population exceeded 50 percent; in Chicago, Los Angeles, St. Louis, Memphis, and other large metropolitan areas, the higher the percentage of nonwhite peoples, the greater the likelihood that a hazardous waste facility existed in the area.[16]

Though they might disagree about whether race/ethnicity or class figured more prominently in the formation of toxic geographies, both Gibbs and the Bullards concurred that marginalized groups overwhelmingly populated the hazardous zone. From Houston to Niagara Falls, hazardous waste disposal seemed to be defined by a borderland mentality that divided landscapes into contained and ultimately acceptable places for toxic dumping. And if you lived there, toxic autobiographers would show in increasing depth and detail over the next several decades, you were not only on the wrong side of the tracks but also often mired in a hazardous existence that impacted one's mind as well as body.

The Toxic Soul

For the first generation of toxic autobiographers, particularly those writing from Love Canal, narrative form usually followed political function. Crafting personal narratives out of a toxic crisis offered grassroots activists a way to connect to a broader movement for environmental justice. Early toxic autobiographies were rarely long (Gibbs's narrative was around 150 pages and was probably the longest such book) and usually not stylized. Many were in fact episodic—comprised of bits and parts (oral testimony or scraps in a personal journal). But many of these early narratives had an additional function beyond grassroots politics: they allowed people to talk about something forbidden. Uncovering a problematic past that few people knew about, or many people hoped to avoid, became a liberating experience. In fact, for some, confronting historical amnesia about a dump's placement, or taking on political officials who ignored citizens' complaints about hazardous waste, became part and parcel of a new as well as more authentic and powerful self. The price of being a truth teller, however, may well have been one's sanity.

Love Canal activists conjured into being the toxic soul—a tormented figure who feels nearly overwhelmed by the responsibilities of reporting on the hazardous waste zone. Jim Clark, a Vietnam veteran and Love Canal activist, told congresspeople that "adverse health effects" were rampant in his neighborhood, but no one really believed him. With little hope for the future (his contaminated home was worthless) and many health officials doubting the nature of chemical hazards in streets further away from the Love Canal dump proper, some residents felt forsaken. Clark knew people who contemplated suicide. "We just discovered another suicide the other day," he ruefully recalled. Marie Pozniak talked about the "confusion and stress" that now shadowed Love Canal activists. Like Clark, she felt that health and political officials doubted many residents' stories—partly out of the fear that a mass evacuation of Love Canal would set an expensive precedent for other communities in the toxic zone. As a "victim" of an unnatural disaster, Pozniak often felt helpless. But like Clark, she also believed that apathy was unacceptable. By telling her story, Pozniak hoped to spur new hazardous waste policies that would ensure "clean air, water, and most importantly, clean and safe homes" for future generations.[17]

While others became activists too, they frequently described a toxic trauma that continued to impact their lives. Patricia Brown, whose home abutted the buried dump, wrote of the alienation and self-doubt that long accompanied her Love Canal life. As she recalled in an unpublished reminiscence, it had been nearly impossible to endure "the nation's first toxic chemical nightmare." Why? "Despite all the publicity" associated with the crisis, she remembered feeling that "no one was there to listen." Worse, some loved ones turned away from victims out of concern for their own well-being. "The world outside seemed to fear us because they thought we were contaminated," Brown wrote. "Friends and family stopped coming to our home because of anxiety for their own safety. We ourselves withdrew from those 'outside' our zone of disaster to protect our self-respect and possessions from offense over which we felt we had no control." Brown felt very much like a "prisoner of forces alien to [my] understanding"—an experiment in psychological endurance. She survived, but not without a variety of psychological scars.[18]

Sociologists studying Love Canal and other chemical disasters note that survivors like Clark, Pozniak, and Brown were right to feel these psychological strains. Because chemical spills and leaking hazardous waste dumps often lack the destructive visual imagery associated with hurricanes, earthquakes, and floods, outsiders can more easily dismiss the nature of the problem. Even victims wonder if they are imagining their illnesses.[19] Having to constantly explain the hidden nature of a toxic problem (as opposed to pointing to a crushed house) thus exerts a unique type of emotional stress; victims often feel under siege. Does anyone believe my story? Hillis's unpublished memoir, "Love Canal's Contamination," offered the most vivid representation of this phenomenon, picturing some residents as on the verge of a nervous breakdown. A young mother and housewife, Hillis had already had a miscarriage and was therefore sensitive to the idea that a toxic problem existed in her neighborhood (health studies showed that area women closest to the dump were 1.5 times more likely to have a miscarriage). When her son and husband became sick, she became enraged. But her complaints led nowhere. State health officials maintained that Hillis lived too far from the Love Canal dump to be affected (and thus moved). In fact, some politicians and health officials labeled Love Canal activists like her "hysterical." Hillis recalled a string of absurd images to prove to herself that she was not crazy: a nearly bald dog who roamed the neighborhood, trees leafing out in July, and acrid chemical smells day and night. She also recalled an insensitive piece of advice from the New York State health commissioner, Davis Axelrod, who told worried residents not to obsess about local chemical hazards because "life is a chancy thing, you know." "God help us," Hillis (n.d.) concluded, because it seemed that few people would lend their support.[20]

Like Brown, Hillis emphasized the division between insiders and outsiders, those who lived in a toxic zone and those who did not understand the dread associated with life there. When she felt sick, Hillis was devastated by her own doctor's dispassionate response. As she noted, he dryly remarked that he was going to watch with interest as "your case develop[s]." Here and on other occasions, Hillis wanted to "scream out" as loudly as she could, but often found that nothing happened; she too was rendered mute by her toxic soul. "I don't want to be a Love Canal victim," conceded Hillis (ibid.), "but oh God, I am"

Writing became a sort of therapy for Hillis. By keeping a scrapbook of notes, correspondence, poetry, and prose—which became the contents of her manuscript—she hoped to translate her anger and anxiety into a sense of purpose. Occasionally, she wrote to her local paper to publicize her thoughts. Indeed, she declared that Love Canal residents had a duty to "scream out loudest of all," and vowed, "I may be sick from chemicals but I'm not *dead*." Joining street protests, testifying in Congress, and confronting health officials, Hillis attempted to take control of her life. At one point, Hillis and others staged a daring act of civil disobedience by refusing to leave local motels after a temporary evacuation order ceased. A fugitive from her toxic neighborhood, she fought her rendition through an explosive sense of rage. Hillis (ibid.) threw knives and a telephone book at one official, screaming that he had no idea what he was condemning her to back at Love Canal.

Unlike Gibbs's autobiography, Hillis's memoir offered no triumphal ending celebrating grassroots environmentalism; it merely stopped with the total evacuation of Love Canal in the early 1980s. Did she ever plan to publish her narrative? Though she did not confront that matter directly,

Hillis seems to imply that her toxic memories of Love Canal remain simply too hazardous to revisit. She had already done her part by speaking in Congress, at rallies, and through letters to the editor. But unlike many other toxic autobiographers, she saw fit only to leave her tale in the archive for future scholars to mine.

More Hazardous Memories: Second-Generation Writers

Throughout the 1980s and early 1990s, toxic autobiography remained part and parcel of rising environmental justice struggles. Writing about a toxic past and poisoned present abetted a sense that grassroots activists were storming the political gates to put toxic waste issues on the map. While these political struggles continued, toxic writing matured as a genre. In fact, by the turn of the century, a new generation of autobiographers emerged to define toxic autobiography's second wave. Often coming from professional or artistic backgrounds (as reporters, writers, and scientists), these authors produced more polished and stylized narratives—almost novelistic in their portrayal of toxic experiences. Differences abounded between the first and second wave. Where earlier authors usually published with university or lesser-known presses (and remained hidebound in terms of audience), many recent writers have had the support of major publishing houses. Moreover, second-wave authors did not have to legitimize hazardous waste as a topic of societal interest; a large body of state and federal laws had long since done that.

Yet while toxic autobiography evolved between the mid-1990s and the present, seeking (and gaining) more literary respect over time, many of its practitioners remain concerned with the themes established by earlier writers. The much-praised *Living Downstream* by Sandra Steingraber (1997), perhaps the most notable toxic autobiography since Gibbs's narrative, uses a similar language of revelation to examine the way that hazardous waste pervaded middle America and suburbia. Like Gibbs, Steingraber once thought that her childhood home of Taswell County, Illinois, was idyllic. Rural and seemingly without the major industries of the rust belt, it could not have seemed farther away from the nation's toxic grid. But, she learned, toxic hot spots emerged from pesticide use and unregulated chemical disposal. Still, her parents' generation said nothing, hoping that any toxic problems would remain buried.

Tragically, she reports, the opposite occurred: Taswell's toxic hot spots claimed more and more victims (usually to cancer) through time. A biologist and cancer survivor herself, Steingraber (ibid., xx) concluded that

there is an undeniable link between forgotten hazardous pasts and the
toxic present—something she confirmed as she traveled around the country. Less than a year after *Living Downstream* debuted, Steingraber noted
in an updated preface, she had "sat down with wheat farmers in Mon-
tana, breast cancer activists in Montréal, and mothers of children with
cancer in New Jersey," not to mention "Massachusetts retirees fighting
toxic dumping in the lakeside homes, Texans fighting contamination of
drinking water with pesticides," and a seemingly endless string of people
at "public hearings" on the environmental incidence of cancer. Turning
these supposedly unconnected events into a national narrative of toxic
life and its discontents, *Living Downstream* showed that Americans from
all places and walks of life now had firsthand experience with toxicity
(ibid., xv–xvi).

Interestingly, soon after Steingraber's book was published, the Environmental Protection Agency (EPA) reported that roughly a quarter of the US
population resided within a half-dozen miles of a Superfund site.[21] This
fact prompted a range of writers to look back and probe their youthful
landscapes for evidence that their parents and communities knew about
the toxic perils that might someday confront them. Nowhere is this theme
better illustrated than in *Body Toxic*, an autobiography about growing
up in cold war New Jersey by Susan Antonetta (2001). Part memoir, part
experimental text (with expressionistic excerpts from her personal journal interrupting the formal narrative), Antonetta's work suggests that
both the US body and mind have become casualties of an unacknowledged toxic past. Like her own family, too many Americans denied the
existence of toxic waste in their midst. She notes that silences enveloped
those around her, even as they become sick. Indeed, the first affliction suffered by the body toxic is the repressed ability to even speak about one's
hazardous past. Though rivers and wooded areas around her home are
saturated with chemical waste, her family did not discuss toxic concerns
until it was too late. When thousands of corroded chemical barrels were
discovered in a nearby field, Antonetta wonders if her parents ever knew
about the unauthorized dump. "Oh yeah," her mother admits. "Everyone
dumped their stuff back there." "How do you know?" Antonetta asks
bluntly. Her mother says that "we used to take walks," but quickly stops
talking about it. She "won't tell me anymore," Antonetta (ibid., 20) grimly
concludes. Although she refuses to blame anyone, she makes it clear that
there is a societal price to those generational silences: an ever-expanding
toxic waste bubble. Referring to the Superfund, Antonetta observes that
New Jersey contains roughly 10 percent of the nation's most toxic sites

(111 out of roughly 1,100)—a staggering number for so small a state. She implies that a society that would produce all this toxic material is probably insane (ibid., 30–31).

Other contemporary works share Antonetta's sense that toxic memory is a hazardous terrain. Kelly McMasters's *Welcome to Shirley: A Memoir from an Atomic Town* and Nancy Nichols's *Lake Effect: Two Sisters and a Town's Toxic Legacy*, both published in 2008, show the various ways that toxic waste has invaded the landscape of formerly pleasant family memory. McMasters's autobiography recounts her "magical childhood" on Long Island's East End during the 1970s and 1980s. Significantly, it is only as an adult looking back that McMaster (2008, xv) sees how nuclear waste ripped apart her hometown. Though Shirley had long been a gritty working-class community, it was founded in the 1930s as "the town of flowers," serving as an oasis from New York City. That bucolic image stayed with the town leaders for much of the century. Indeed, as nuclear power plants and research facilities such as Brookhaven National Laboratory invaded the area, few people worried about the impact of mass industrialization. It was more important to live the American dream and win the cold war.

Later, when family and friends became ill with various cancers, McMasters begins to wonder about her beloved hometown's toxic past. "While I was growing up," she explains, "Brookhaven . . . was a fixture of my imagination. . . . Much like the proverbial monster in the basement, the lab obsessed us precisely because it was so close." An economic engine for the community, it seemed larger than life. But what actually happened inside? And how did that impact the surrounding landscape? McMasters shows that denial and deflection, rather than a full reckoning with these problematic questions, proved easier for politicians and local residents alike. "We had no access [to Brookhaven]," she remembers. "We couldn't see the buildings or the scientists who populated them." Researchers lived far away from Shirley, while the "neighborhood fathers [technicians and service staff who] . . . spent their days there refused to talk much about what their jobs actually entailed." And no one "ever discussed evacuation plans or the possibility of a nuclear meltdown so close to home" (ibid., 94–95).

Like Antonetta, McMasters does not castigate anyone for this lack of knowledge, though her quest to retell the town's toxic past indicates uneasiness with the mantra "that's just the way things were." She satirizes Shirley's attempt at forced amnesia in the 1980s. After details of a nuclear leak surface, the town leaders seek a new start by holding a name-changing contest. Even though a technicality prevents the new name from taking hold, McMasters is stunned that there is no public discussion of Brookhaven's potential health hazards on Shirley. Like the nearby nuclear facilities, everything is hidden in plain sight. It is up to McMasters to uncover and remap the toxic town she still loves.

McMasters's story of toxic health struggles continues virtually up to the present. She finds the disconnection between ever-more revelations of radioactive waste (in dumps, the drinking water, and a nearby river polluted beyond recognition) and the upbeat responses from Brookhaven scientists and political officials nothing short of numbing. On one street bordering Brookhaven, colloquially known as Death Row, McMasters visits residents who form a protest bloc that eventually launches a billion-dollar class action lawsuit against the lab's management company. Significantly, McMasters writes, Love Canal lawyers headed the ongoing case. McMasters finally admits that as a young adult, she longed to be anywhere but Shirley. After attending school in Upstate New York, marrying, and settling into a life far from Shirley, she can only look back with sadness. *Welcome to Shirley* rememorizes the town she cannot forget—literally so, as concerns about toxins in her family and body linger for years.

In *Lake Effect*, Nancy Nichols provides a less wistful but no less devastating portrait of Waukegan, Illinois. Nichols's narrative has a funereal quality, flowing as it does from a promise she made to her dying sister, Sue, to investigate the poisoned landscape of their youth in a seemingly wonderful midwestern locale. Though they grew up happy in this nearly picture-perfect town on the western edge of Lake Michigan—the birthplace of sci-fi writer Ray Bradbury, who immortalized it in his 1957 novel, *Dandelion Wine*, as a beautiful "Green Town"—both were diagnosed with cancer as adults. Before passing away, Sue compels Nancy to reexamine the toxic legacy that engulfed them. "You have to write about this," Sue pleads. "I will," Nancy, a reporter by trade, replies, realizing that this assurance was nothing less than a call to "make meaning out of her [sister's] agony" (Nichols 2008, 4).

The toxic history that Nichols uncovers stretches back decades, prompting a series of reflections on her polluted childhood. New reports from the 1970s onward about environmental degradation in Waukegan indicated that so-called Green Town might not be the ideal landscape that the city's leaders once claimed. Both Johns Manville, an asbestos producer, and Outboard Marine Corporation, a manufacturer of PCBs, operated plants along the waterfront for years. Yet the extent of their toxic trail was not fully revealed until much later. Like the howling winds that reach well beyond Lake Michigan, these toxic revelations will impact a great many people down the road, and as Nichols notes, "play a role in my family too" (ibid., xi).

Vivid chapter titles carry readers along Nichols's quest to remap her childhood landscape. "Green Town" examines the industrial history that shaped the town, including a rusty aftermath that filters through Nichols's family; "Lake Michigan Legacy" meditates on her son's future in a toxic world; "Miasma" refers to both the putrid smells associated with Waukegan's toxic harbor and "a kind of information miasma" that perverted local consideration of the town's industrial past—a past that company officials elided by removing "many documents and artifacts" about chemical pollution from the local historical society. In "The Used Car Salesman's Daughters," Nichols continues to meditate on the theme of buried truths by linking her father's disingenuousness (he was a used car salesman) to Waukegan's approach to environmental pollution. In 1984, she reports, after a corrupt deal between town officials and an EPA representative was uncovered, "the full extent of Waukegan's chemical contamination was revealed": a million pounds of PCB contamination in the harbor. "Just about every chemical known to be dangerous to human health is in one of those sites," one expert tells her. Although three Superfund sites have been created, most area residents are left to determine how they might have been impacted through the years (ibid., 35, 47, 67, 46).

By the end of the book—in a chapter titled "Proof"—Nichols strikes an outraged tone. While there is more than enough scientific evidence to connect toxic environments and ill health, many state and federal courts often side with chemical manufacturers. "It is nearly impossible to prove that a particular person's cancer results from a particular chemical," Nichols notes, "at least in terms of the standards of legal proof." Her toxic tour of Waukegan's past becomes all the more important, then, creating a narrative warning system about the day-to-day landscape in which she lived. And yet, she continues, many people would rather ignore the lessons of a poisonous past. Nichols shows that new development along the Waukegan waterfront fails to account for hazardous waste abatement. The "Waukegan story," she concludes rather sadly after looking at other hazardous waste sites in the United States, is not unique" (ibid., 128, 135, 35, 47).

Toxic Literature and Green Knowledge

After decades of practice, just where does toxic literature fit within contemporary environmental thought and activism? In one key sense, toxic literature has brought environmentalism back to literary outreach

(Cronon, 1992). As Daniel Payne has contended, US environmental writing ascended as a genre during the nineteenth and early twentieth centuries precisely because conservationists and preservationists sought to mobilize readers on behalf of nature. From John Muir to Teddy Roosevelt, early environmentalists claimed "the rhetorical high ground" by investing the US public in the very cause of environmental reform. In Payne's eyes, environmental literature—stories about endangered nature—helped build the early environmental movement. Yet while the modern environmentalism of the 1960s and 1970s initially cultivated widespread public support of ecology, its focus on institutional reform soon moved environmental debate into the more specialized realms of law and technoscience. "The most significant recent development affecting the role of nature writers in environmental reform," Payne observes, "may well be the fact that the battle for environmental protection now often takes place in the courtroom." Though yielding important results, this legal/scientific turn has also marginalized environmental writing and writers (Payne 1996, 167–75). Toxic literature seeks to return environmentalism to its consciousness-raising past.

Indeed, while environmental justice movements rely on legal and scientific expertise, toxic literature remains striking for its continued emphasis on broad public outreach—on claiming the moral high ground over hazardous waste concerns. From its earliest incarnations at Love Canal to recent ruminations on toxic communities nationally and globally, toxic literature has argued that there is a vital human dimension to hazardous waste policies and procedures. How we handle toxic waste is not, in other words, a technical or legal issue best left to experts. Rather, it is part of a society-wide moral debate about environmental winners and losers. Who must embody toxic troubles, and why? As generations of toxic autobiographers have pointed out, the concept of environmental risk is not abstract; rather, it is highly personal and communal, accounted for in stories of hazardous geographies, biographies, and histories. From Love Canal to Waukegan, someone, somewhere, must live with the public health consequences of toxic waste.[22] In this sense, toxic autobiography has also pushed environmental literature forward. For stories about embattled nature in Muir's time have become narratives about endangered human nature in Gibbs's day.

Looking at the global perspective further underscores the point. In communities from Africa to India, local people have begun to tell stories about the way that industrial and technical modernity has threatened as much as improved their lives.[23] If toxic concerns form only a subset of these narratives—other concerns include soil erosion, the clear-cutting of forests, and massive energy projects that displace traditional communities—they are nonetheless a foundational part of a global green front. In fact, as Adriana Petryna has shown, global toxic tragedies have given rise to the concept of "biologic citizenship"—the notion that citizens in postindustrial society must organize around environmental health grievances. Studying Chernobyl, Petryna (2002, 5) found that former nuclear workers as well as survivors used their toxic experiences to innovate new concepts of both "social membership" and civic identity. For them, as for those living in America's toxic grid, all modern nation-states had a duty to define and protect citizens' environmental rights.

And make no mistake: toxic autobiography has impacted US politics and government. From South Central Los Angeles to Long Island, the consciousness-raising efforts of toxic writers helped uncover poisoned pasts and remap the meaning of the nation's toxic grid. Stories about environmental ruin in fact compelled state and federal politicians to create much of the legislation defining contemporary toxic policy, from the Superfund (the nation's first hazardous waste cleanup law, which since its inception in December 1980 has mandated remediation of the nation's worst hazardous waste sites, including Love Canal) to state/federal "right-to-know" laws (which allow local residents to access information on hazardous waste emissions in their neighborhood). As Senator Ed Muskie of Maine, one of the leading congressional advocates of environmental protection, put it in supporting the Superfund, the testimony of Love Canal women convinced him that more must be done to protect citizens from the threat of toxins in their own communities.[24]

As Muskie's words indicate, women have a played a key role in toxic autobiography's rise as a genre. They have transformed definitions of both environmental activism and environmental space. Using their day-to-day domestic experiences with hazardous waste, women have reenvisioned the US environment itself as a localized space where neighborhoods, homes, schools, and playgrounds constitute a threatened (and often-threatening) landscape. Nature is not out there, in other words, but where we live. Whether in the urban core, on Native American lands, in the rural United States, or in emerging suburbs, local women have frequently been the first to complain about toxic pollution in their midst. Often marginalized politically and socially, they have become visible via personal narratives seeking some form of environmental justice.

In this manner, toxic literature may ultimately meet Raymond Williams's famous definition of an "emergent" cultural form. Williams (1977,

123) argued that dominant value systems regularly spawn oppositional cultures, some of which are "merely novel" (and thus more easily absorbed into mainstream culture), and others that engender "new meanings and values, new practices, new relationships." Though some environmental theorists still relegate environmental justice advocates to the outer realm of "green knowledge"—that is, they signify the merely novel, not substantive, critiques of mainstream values—even a brief survey of toxic literature reveals that its practitioners have long been concerned with challenging fundamental environmental values in the United States. By emphasizing just who has usually occupied the toxic grid—marginalized socioeconomic and ethnic groups—generations of toxic writers have shown that US hazardous waste policies have targeted certain groups and places as the equivalent of the "deserving poor." And even in more recent cases where toxic literature has reached into suburbia and/or the US heartland, new waves of autobiographers and investigative reporters seek to illustrate that hazardous waste flows from a problematic, often-unacknowledged past that valorized industrial and high-tech development above human and environmental safety. Only by opening up that toxic past, and investigating the seemingly inordinate power of science and industry to frame debates about hazardous waste in terms of technological progress, economic efficiency, and so-called nonbiased public policy (rather than those revolving around ecological principles and/or environmental justice), can Americans better regulate their local as well as national environments.[25]

But perhaps toxic literature has done something more basic. By making hazardous waste a key component of modern environmentalism, toxic writers defined average citizens as environmentalists. For that reason alone, toxic literature may one day assume global, not just local, importance.

Notes

1. *Navajo Times*, July 2009.

2. Hillis n.d.; pages not numbered consistently in original.

3. See Buell 2005, 17–24.

4. On Carson, see especially Lear 1997.

5. See, for example, the comments of environmentalist A. Blackman Early, who worked for Environmental Action, on delayed clean water regulations in the *New York Times*, ("Ecologists Disagree on Clean-Water Bill," November 20, 1977.

6. See Newman 2001.

7. See Brown 1979, 1981.

8. See Morell and Margorian 1982 1–8.

9. *The Killing Ground* was updated on August 21, 1980.

10. See also Newman 2001.

11. For an excellent analysis of Love Canal's various activist groups, see Blum 2011.

12. Ibid., 8–9.

13. See Breton 1998, 128.

14. Grinde and Johansen 1995, 203.

15. Ibid., 90–92.

16. See "Toxic Waste and Race in the United States," reprinted in Payne and Newman 2005, ix, xi.

17. For Clark's story, see Hillis n.d. For Pozniak's recollections, see Marie Pozniak, Love Canal Resident 1979.

18. For Brown's reminiscences of Love Canal life, see Patricia Brown Papers n.d.

19. See especially Levine 1982, 1–2.

20. For an early but still relevant sociological perspective, see Levine 1982.

21. For an interesting history of the toxic grid, see http://www.epa.gov/superfund/sites/index.htm.

22. On environmental risk and the problem of toxic knowledge in a risk society, see Beck 1992. Although this work makes critical errors on Love Canal, for an opposing view, see Sunstein 2004.

23. On global green struggles in Africa, see, for instance, Matthei 2006. On India, see also Shiva 1989.

24. On Love Canal women's impact on politics, see Newman 2001.

25. On the matter of community environmentalists and green knowledge, see Jamison 2001, 147–15.

References

Antonetta, Susan. 2001. *Body Toxic: An Environmental Memoir.* Washington, DC: Counterpoint.

Beck, Ulrich. 1992. *Towards a New Modernity.* Thousand Oaks, CA: Sage.

Blum, Elizabeth. 2011. *Love Canal Revisited.* Lawrence, KS: University Press of Kansas.

Breton, Mary Joy. 1998. Cathy Hinds . . . Battles Poison in Maine. In *Women Pioneers for the Environment*, ed. Mary Joy Breton, 125–132. Boston: Northeastern University Press.

Brown, Michael H. 1979. Love Canal and the Poisoning of America. *Atlantic Monthly* 244 (6): 33–47.

Brown, Michael H. 1981. *Laying Waste: The Poisoning of America by Toxic Chemicals*. New York: Washington Square Press.

Buell, Lawrence. 2005. *The Future of Environmental Criticism*. Malden, MA: Blackwell.

Bullard, Robert. 1990. *Dumping in Dixie: Race, Class, and Environmental Quality*. Boulder, CO: Westview Press.

Carson, Rachel. 1962. *Silent Spring*. New York: Houghton Mifflin.

Citizens' Clearinghouse for Hazardous Waste. 1983. Leadership Handbook on Hazardous Waste. Arlington, VA: Citizens' Clearinghouse for Hazardous Waste.

Cronon, William. 1992. A Place for Stories: Nature, History, and Narrative. *Journal of American History* 78, no. 4 (March): 1347–1376.

Fadiman, Clifton, and Jean White, eds. 1971. *Ecocide . . . and Thoughts toward Survival*. New York: Interbook.

Gibbs, Lois. 1982. *Love Canal: My Story*. Albany: State University of New York Press.

Gibbs, Lois. 1998. *Love Canal: The Story Continues*. Gabriola Island, BC: New Society Publishers.

Grinde, Donald A., and Bruce E. Johansen. 1995. *Ecocide of Native America: Environmental Destruction of Indian Lands and Peoples*. Sante Fe, NM: Clear Light Press.

Hamilton, Cynthia. 2000. Women, Home, and Community: The Struggle in an Urban Environment. In *A Forest of Voices*, ed. Chris Anderson and Lex Runciman, 542–546. New York: McGraw-Hill.

Hillis, Anne. n.d. Love Canal's Contamination: The Poisoning of an American Family. Unpublished manuscript. Papers of the Ecumenical Task Force of the Niagara Frontier, State University of New York at Buffalo, Love Canal Collections, James L. Clark Testimony" before the Senate Committee on Resource Protection, the Environment, and Public Works. http://library.buffalo.edu/specialcollections/lovecanal/collections/etf.php

Jamison, Andrew. 2001. *The Making of Green Knowledge: Environmental Politics and Cultural Transformation*. New York: Cambridge University Press.

Jones, Charles O. 1972. From Gold to Garbage: A Bibliographical Essay on Politics and the Environment. *American Political Science Review* 66, no. 2 (June): 588–595.

Kiefer, Irene. 1981. *Poisoned Land: The Problem of Hazardous Waste*. New York: Atheneum.

Kenny, Luella. 1979. Love Canal Testimony Submitted to the House Subcommittee on Oversight and Investigations by Luella Kenny. April 2. http://library.buffalo.edu/specialcollections/lovecanal/collections/etf.php.

The Killing Ground. 1979. ABC News, March 29.

Lear, Linda. 1997. *Rachel Carson: Witness for Nature*. New York: Henry Holt.

Levine, Adeline. 1982. *Love Canal: Science, Politics, and People*. Lexington, MA: D. C. Heath.

McCoulf, Grace. 1979. Testimony, April 5. http://library.buffalo.edu/specialcollections/lovecanal/collections/etf.php

McMasters, Kelly. 2008. *Welcome to Shirley: A Memoir from an Atomic Town.* New York: Public Affairs.

Matthei, Wangari. 2006. *The Greenbelt Movement: Sharing the Approach and the Experience.* New York: Lantern Books.

Morell, David, and Christopher Margorian. 1982. *Siting Hazardous Waste Facilities: Local Opposition and the Myth of Pre-Emption.* Cambridge, MA: Ballinger Publishing Company.

Newman, Richard. 2001. Making Environmental Politics: Women and Love Canal Activism. *Women's Studies Quarterly* 29, nos. 1–2 (Spring–Summer): 65–84.

Nichols, Nancy. 2008. *Lake Effect: Two Sisters and a Town's Toxic Legacy.* Washington, DC: Island Press.

Patricia Brown Papers. n.d. Papers of the Ecumenical Task Force of the Niagara Frontier, State University of New York at Buffalo, Love Canal Collection, box 1, folder 1.

Payne, Daniel G. 1996. *Voices in the Wilderness: American Nature Writing and Environmental Politics.* Hanover, NH: University Press of New England.

Payne, Daniel G., and Richard Newman. 2005. *The Palgrave Environmental Reader.* New York: Palgrave.

Petryna, Adriana. 2002. *Life Exposed: Biological Citizens after Chernobyl.* Princeton, NJ: Princeton University Press.

Pozniak, Marie (Love Canal Resident). 1979. Testimony before the US Senate Subcommittee on Toxic Substances and Chemical Wastes, May 3. http://library.buffalo.edu/specialcollections/lovecanal/collections/etf.php

Reed, T. V. 2009. Toxic Colonialism, Environmental Justice, and Native Resistance in Silko's *Almanac of the Dead. Melus* 34, no. 2 (June 2009): 25–42.

Rodale, J. I. 1964. *Our Poisoned Earth and Sky.* Emmaus, PA: Rodale Books.

Shebala, Marley. 2009. Poison in the Earth: 1979 Church Rock Spill a Symbol for Uranium Dangers. *The Navajo Times Online* (Window Rock, AZ), July 23. http://navajotimes.com/news/2009/0709/072309uranium.php.

Shiva, Vandana. 1989. *Staying Alive: Women, Ecology, and Development.* London: Zed Books.

Steingraber, Sandra. 1997. *Living Downstream.* New York: Vintage.

Sunstein, Cass. 2004. *Risk and Reason: Safety, Law, and the Environment.* Cambridge: Cambridge University Press.

Williams, Raymond. 1977. *Marxism and Literature.* New York: Oxford University Press.

2

"The Most Radical View of the Whole Subject": George E. Waring Jr., Domestic Waste, and Women's Rights

William Gleason

Historians of the US obsession with cleanliness might well find in the popular 1885 advice manual *How to Drain a House: Practical Information for Householders* by George E. Waring Jr. ample evidence of the near hysteria that at times accompanied calls for improved domestic hygiene in the late nineteenth century. From the handbook's prefatory remarks, ominously titled "Our Enemy the Drains"—in which Waring (1885, iii) declares that "the drains in average modern houses are probably the most serious and prevalent enemies with which struggling humanity has to contend"—to the volume's repeated insinuation that dirt and decay threaten to undermine civilization itself, *How to Drain a House* often reads more like a combat manual than a do-it-yourself guidebook.

Moreover, Waring's insistence that it is to middle-class US housewives that the burden of such warrior duties fall would seem to make his text complicit with an array of late-century cultural forces that simultaneously created and required, in the incisive formulation by Ruth Cowan Schwartz (1983), "more work for mother." Indeed, women readers hoping to escape what might well have appeared to be drudgery rather than glory would have been duly chastened by the closing paragraph of Waring's preface, in which he spells out the weighty duties of the "woman of the house." "There has been placed under her control," Waring (1885, v–vi) advises,

a means of safety, or an engine of destruction, according as she performs her duty, or neglects it. She can not safely delegate her responsibility to her servants. Her own eye must see that at no point, has neglect, at any time, permitted even the beginning of filth—for the beginning of filth is the beginning of danger. It marks the desertion of the ally to the ranks of the enemy.

This isn't just a call to clean; Waring's preface instead places the housewife at the head of an intensely surveillant regime. Even the servants must be watched, as the very house must be watched, for the smallest sign of neglect—for "even the beginning of filth"—lest imminent "destruction" ensue. Few late-century texts, one might fairly conclude, sutured women's duties to apocalyptic anxiety about dirt and waste as effectively as Waring's.[1]

How fascinating, then, to discover that *How to Drain a House* stands at the end of a phase in Waring's career during which the well-known drainage and sewerage specialist frequently appeared far less certain about the roles that US women should play in the broader fight for sanitary reform. If we return to the beginning of this phase, in the early 1870s, for example, we find Waring imagining women not as housebound hygienists but rather as environmental activists, inspired to remediate an increasingly degraded public landscape through recognition of their own sociocultural subordination. Thus, as opposed to simply condemning Waring for the chauvinistic myopia of *How to Drain a House*, we can trace the contours of an ongoing cultural conversation, in which Waring was a catalytic participant, about gender and garbage, waste and technology, and even women's rights and national character. In particular, Waring's provocative call for women to take control of their own sanitary fate put women's health issues, for a time, at the forefront of discussions of US privy reform, a pressing environmental concern of the late nineteenth century—although one that other commentators were far less eager to envision as the basis for a radical reimagining of women's environmental rights.

The participants in this conversation included not just "sanitary engineers" (like Waring) but also doctors, public officials, and journalists. As we will see, during the 1870s and 1880s, two decades that marked the US sanitary reform movement's coming of age, it was just as likely for a gynecologist—or the US commissioner of agriculture—to make public pronouncements about women, waste, and public health as for a sanitary specialist like Waring to do so. Many of these declarations concerned themselves intimately with the waste products of human bodies, the excrement and urine whose efficient disposal was rapidly becoming the central problem of the new US city. By linking the importance of waste management to women's health early in his career, Waring demonstrates an acute awareness that environmental management was not merely a technological problem but also a social and cultural one. Arguing vigorously, for example, that women are less likely than men to use poorly designed, distant, or exposed privies (and so are more likely to forego the daily exercises of evacuation so essential to physical well-being), Waring lays as much stress on changing political roles (men's and women's) as he does on upgrading traps and pipes. Although Waring's excessive concern with the regularity of women's bowels might seem prurient or patriarchal, he shows not only a striking sensitivity to the nontechnological barriers that might compromise women's health but also—at least initially—a powerful sense of the ways women themselves might take public action to effect change.

In the pages that follow I reexamine Waring's early writings on commodes, drains, and sewers in relation to parallel discussions of women and waste in government reports, sanitary periodicals, and medical lectures. This period of Waring's career is important to reconsider in part because the spotlight that environmental historians have justly turned on his triumphant success as the street-cleaning commissioner of New York in the mid-1890s—where he not only devised the infamous "White Wings" trash collection brigade but also introduced large-scale recycling and source separation to the largest city in the country—has inadvertently thrown this earlier phase into obscuring shadow. In tracing Waring's underappreciated endorsement of women's environmental activism along with his own gradual retreat from the most radical implications of this position, we can also see more clearly how the debates over waste control in the 1870s and 1880s both temporarily unsettled and ultimately reproduced dominant US gender relations.[2]

Water, Waste, and Rights

Waring's broad interest in questions of drainage and waste matter can be traced to his first book, *The Elements of Agriculture: A Book for Young Farmers, with Questions Prepared for the Use of Schools*. Written when Waring was only twenty, the product of his training in the early 1850s with the scientific agriculturalist James J. Mapes, to whom the volume is dedicated, *The Elements of Agriculture* follows three main sections on plants, soil, and manure (both animal and human) with a fourth on "mechanical cultivation," which includes two chapters on the proper methods for draining sodden fields. A strong proponent of underdraining, in which drains of burned clay pipe or loose stone are placed beneath the ground rather than left in open furrows, Waring offers his readers detailed instructions for "cheaply, thoroughly, and permanently" draining land. In addition to improving the fertility and profitability of US farms, properly drained land, Waring opined, might supply additional benefits for public health. "One very important, though not strictly agricultural, effect of thorough drainage," he notes in a short paragraph, "is its removal of certain local diseases, peculiar to the vicinity of marshy or low moist soils. The health-reports in several places in England, show that where *fever and ague* was once common, it has almost entirely disappeared since the general use of under-drains in those localities" (Waring 1854, 216, 231; emphasis in original).[3] Although Waring's stubborn adherence to a miasmatic model of contagion would later bring him into conflict with proponents of germ theory, discrediting him in the eyes of some critics, in this brief allusion to the health benefits of drainage we can see the seeds of what in due time would become Waring's consuming obsession: how best to manage the wastes of houses and towns.

Within just a few years, Waring would become the drainage supervisor on one of the most high-profile public works projects in the United States. After parlaying the success of *The Elements of Agriculture* (to which his publisher, D. Appleton, quickly announced a planned sequel—though none ever appeared) into a position as the agricultural manager of *New York Tribune* editor Horace Greeley's suburban Chappaqua farm, in 1857 Waring would be hired by Frederick Law Olmsted to oversee the drainage of Manhattan's new Central Park.[4] (Waring had entered the public competition to design Central Park, and although his entry was not selected, Olmsted was impressed by the thoroughness of his report on the drainage requirements.) Working for the Park Commission until the outbreak of the Civil War in 1861, Waring helped transform acres of swampy wetland into usable public space with the aid of over sixty miles of strategically buried clay pipes, some of which still carry water today.[5]

After the war, and while serving as manager of the experimental Ogden Farm near Newport, Rhode Island, Waring turned his pen to the series of publications that would bring him to the forefront of public discussions over the management of water and waste in the 1870s and 1880s. The first of these volumes was his detailed study *Draining for Profit, and Draining for Health*, published in 1867. Advertised as being written by the "Engineer of the Drainage of the Central Park, New-York," *Draining for Profit* is essentially a book-length expansion of Waring's two-chapter foray into drainage in *The Elements of Agriculture*—reissued as a second edition in 1868—amplified by accounts of his work on Olmsted's park. Brooking no "half-way measures," Waring (1867, 3) announces in the preface to *Draining for Profit* that he "has purposely taken the most radical view of the whole subject," endeavoring "to emphasize the necessity for the utmost thoroughness in all draining operations, from the first staking of the lines to the final filling-in of the ditches." True to his word, he leaves nothing out. Readers are not merely advised to use clay pipes; they are guided through the process of actually making the tiles themselves, from obtaining the proper mixture of sand and clay, to the subsequent tempering, molding, carrying, drying, rolling, and finally burning of the finished product. There are even instructions for building your own kiln. The entire volume is liberally illustrated, featuring nearly fifty woodblock engravings.

The final two chapters of *Draining for Profit*, which focus on the benefits to public health from improved drainage and waste removal, extrapolate Waring's one-paragraph reflection on these topics in *The Elements of Agriculture*. While the first of these chapters, "Malarial Diseases," largely reiterates the arguments Waring had made in the earlier volume in favor of the draining of miasmal land near urban and suburban settlements, the second chapter, "House Drainage and Town Sewerage in Their Relations to the Public Health," introduces a new subject: "the branch of the Art of Drainage which relates to the removal of the fecal and other refuse wastes of the population of towns." This "branch" has an agricultural application, to be sure, in the recovery of valuable human manure for crop fertilization. But its primary aim, like the draining of marshlands, is the prevention of "fatal epidemics" believed to be caused in urban centers like New York by the "vast accumulations of filth, which ebb and flow in many of the larger sewers, with each change of the tide" (Waring 1867, 224, 227).

In this second concluding chapter, Waring stumps for the citywide replacement of large and porous brick collecting drains with smaller and impermeable glazed stoneware pipe sewers—a "radical change" proven effective in England in channeling the discharge and reducing the escape of foul-smelling, disease-ridden effluent. Anything less marks an unnecessary capitulation to the twin evils of bad health and bad husbandry. "The principles herein set forth," he observes, "are no less applicable in America than elsewhere; and the more general adoption of improved house drainage and sewerage, and of the use of sewage matters in agriculture, would add to the health and prosperity of its people, and would indicate a great advance in civilization" (ibid., 228, 239). In short, there is no good reason for your city—or your house—to be fouled (or impoverished) by waste.

In Waring's next publication, the forty-eight-page pamphlet *Earth-Closets: How to Make Them and How to Use Them*, published in 1868, this dictum takes on an even more intimate resonance. Moving directly inside the US home—and using his own dry-earth commode as example and evidence—Waring tackled the sticky problem of waste removal at its human source. Although to this point in his career Waring's work on agriculture and drainage had not yet been linked to the concerns of midcentury US women's household management, this pamphlet would bring him to the attention of two of the field's leading voices: sisters Catharine E. Beecher and Harriet Beecher Stowe.

Persuaded by Waring's passionate endorsement of the earth closet as a cheaper, more healthful, and less labor intensive alternative to the water closet (after using the earth closet, one simply poured a cup of dry earth on the feces or urine to absorb the odor; no running water or pipes were necessary), Beecher and Stowe (1869) devoted an entire chapter of their popular treatise, *The American Woman's Home*, to the system's virtues, quoting at length from Waring's pamphlet and using many of his own illustrations showing women how to build their own commodes. Sanitation historians have tended to treat Waring's detour into earth closets as an example of either his faddishness or his subordination of environmental principles to business interests. Indeed, Waring did have a financial stake in the Hartford-based Earth-Closet Company, the US distributor of earth-system commodes, leading him to promote the device even after it became clear that water closets were being far more widely adopted. And yet the Beechers' embrace of Waring may have played an unacknowledged role in the startling announcement in his follow-up pamphlet, *Earth-Closets and Earth Sewage*, published in 1870, of the concept of a "woman's right" to safe and sanitary home waste management (Waring 1870, 27).[6]

For whereas *Earth-Closets: How to Make Them and How to Use Them* focuses on the somewhat mundane details of construction and disposition—the most dramatic moment perhaps being Waring's announcement (1868, 6) early in the pamphlet that even when the waste pan in his home model "is entirely full, with the accumulation of a week's use," visitors "invariably say, with some surprise, 'You don't mean that this particular one has been used!'"—*Earth-Closets and Earth Sewage* not only conveys a heightened sense of urgency behind the need for a dry-earth waste control system but also offers it as an explicit ameliorative for the physical and social degradations suffered by women condemned to use the typical privies of midcentury America.[7] In fact it is in this latter pamphlet, we might say, that Waring will provide—in political rather than engineering terms—a "most radical view" of the subject of women and waste by articulating for the first time in the literature on privies and health a specific woman's right to clean, safe, and private toilet accommodations (Waring 1867, 3).

In the 1870 pamphlet, Waring's rhetoric rises from indignation to a specific articulation of environmental rights. He first assails the common privy as a source of not merely indecency and discomfort but more specifically women's ill health. That "out-of-door privies, those temples of defame and graves of decency, that disfigure almost every country home in America, and raise their suggestive heads above the garden-walls of elegant town-houses," are "objectionable on the score of decency and comfort, will be confessed by all," Waring writes. "What is not so generally understood," he elaborates, "is their pernicious effect upon health. The influence of subterranean stores of fæcal matter in the propagation of disease has already been referred to, . . . but that which produces, in the aggregate, far worse results—the aggravation of the difficulties of delicate females—has attracted less attention than its importance deserves" (Waring 1870, 23, 26).

After describing, with the help of medical testimony, the injurious effects of bowel irregularity on "the prevailing female complaints," Waring introduces a striking narrative set piece, painting a vivid picture of the special obstacles rural and suburban women face simply in trying to go to the bathroom in an era before widespread indoor plumbing:

Let us see what chance a woman living in the country has to escape the direst evils that "delicate health" has in store for its victims. The privy stands, perhaps, at the bottom of the garden, fifty yards from the house, approached by a walk bordered by long grass, which is always wet except during the sunny part of the day, overhung by shrubbery and vines, which are often dripping with wet, and exposed frequently to the public gaze. In winter, snowdrifts block the way, and during rain there is no shelter from any side. The house itself is fearfully cold, if not drifted half-full with snow or flooded with rain. A woman who is comfortably housed during stormy weather will, if it is possible, postpone for days together the dreadful necessity for exposure that such circumstances require. If the walk is exposed to a neighboring work-shop window, the visit will probably be put off until dusk. In either case, no amount of reasoning will convince a woman that it is her duty, for the sake of preventing troubles of which she is yet ignorant, to expose herself to the danger, the discomfort, and the annoyance that regularity under such circumstances implies. (ibid., 26–27)

Waring's stress on the hardship for women of "exposure," mentioned in some form four times in this paragraph, and meant to encompass both the physical assault of poor weather and the social assault of the intrusive male glance, heightens the paradoxically claustrophobic publicity attendant, for women, on this act. In Waring's imagined scene, the woman cannot avoid being touched by wet grass or dripped on by wet vines on her way to the privy. She has no clear path, and yet her own walk is clearly open to the "public gaze" and especially the men who presumably watch her through the "neighboring work-shop window." Only under the cover of darkness, Waring suggests—if then—will the woman make the trip at all.

It is at this point that Waring's commentary turns political. "Every consideration of humanity, and of the welfare not only of our own families, but of the whole community, demands a speedy reform of this abuse." More specifically, declares Waring,

this suggests a "woman's right," whose acquisition is more vital to her health and happiness than any that the supposed-to-be-coveted suffrage promises her; and she may, with just cause, insist that, however much she may be tyrannized over in the important matters of employment and voting, mankind has no right to hold her longer in subjection to this practical curse. It is hardly more important that she have a house to shelter her from the weather than that this incentive to a dangerous irregularity be removed. (ibid., 27)

Although Waring makes no mention in *Earth-Closets and Earth Sewage* of the Beechers, nor alludes to their endorsement of his system, it seems likely that his decision to present the device as having profound political interest for US women may have been influenced by this contact. And although one might question the extent of Waring's endorsement of women's rights more broadly given his implicit criticism of "the supposed-to-be-coveted suffrage" in the passage above, he closes this section of the pamphlet by calling on women to become political and environmental actors rather than wait for someone else to do this work for them. "In view of the foregoing facts," he notes, "I make no apology for calling the attention of women themselves to this important matter, believing that they will universally concede that . . . their mode of life is neither decent, civilized, nor safe, unless they are provided with the conveniences that the water-closet and the earth-closet alone make possible." (Waring was willing to promote the water closet if it improved women's health, even if in 1870 he still believed the earth closet would become the universal mode of domestic sanitation in all but the most densely populated urban areas. The principal advantage of each model is that it can be situated inside the house, thus eliminating the dangers of exposure detailed above.) Concludes Waring: "Being the parties most interested, it rests with [women] to secure the necessary relief" by taking the initiative to either replace their detached privies with indoor models or convert them, following his step-by-step instructions, into more sanitary outdoor earth closets (ibid., 27). The Beechers could not have said it better.

Unspeakable Relief

That Waring was finding an appreciative—and national—audience among women readers seems clear not just by virtue of the endorsement of the influential Beecher sisters but also by the appearance, in May 1870, of a notice for *Earth-Closets and Earth Sewage* in one of the era's thriving children's magazines, the Chicago-based *Little Corporal.* "This is a neatly-printed, well-written pamphlet of 100 pages, full of valuable information to all who desire to have comfortable, neat, and healthy homes," the notice announces. "All women should be especially interested in this subject" (Book Notices 1870, 77).

By the end of that same year Waring's views, if not always with explicit attribution, were being widely circulated within the North American medical establishment among physicians specializing in the treatment of women. Horatio R. Storer, for example, founder and secretary of the Gynecological Society of Boston (the oldest US gynecologic society), editorialized in the October 1870 issue of the Society's *Journal* on "the importance, in a gynæcological light," of improving the state of public and private "latrinæ." Directing his readers to "the literature of the 'dry-earth system,' at present so deservedly attracting public attention," Storer highlights "a page or two directly to the point as specially affecting the health of women." Storer then quotes at length from what turns out to be the text of an advertisement for Waring's Earth-Closet Company (1869), promising women "unspeakable relief" from repulsive privy accommodations that had appeared with regularity in US newspapers during the preceding year.[8] Although Storer does not name Waring as the author (a virtual certainty, given Waring's position as the Earth-Closet Company's lead spokesperson), the rhetorical similarity to the central narrative set piece of *Earth-Closets and Earth Sewage* is unmistakable:

Probably, ninety-nine out of every hundred habitations in the whole country have nothing better than an unsightly privy, standing at some distance from the house,—too often barbarously foul,—and generally unapproachable except by an entirely unprotected walk, that is more or less exposed to public view, and, in wet or cold weather, is passable only at the risk of getting wet feet, draggling through wet grass or weeds, plodding through snow, or facing cold winds and storms.

As a natural consequence, delicate women soon school themselves to a postponement of the demands of nature, sometimes for days together, rather than expose themselves to the danger of taking cold and to the certainty of great annoyance. Sometimes modesty, and sometimes the dread of discomfort and exposure, is the motive. In all cases the result is the same. The natural functions become disordered, the digestion is impaired, and dyspepsia, with its thousand-and-one horrors, breaks down the constitution and lays the foundation for all manner of "female complaints." (Storer 1870, 271–72)

In one form or another—and often with increasingly hyperbolic stress laid on the vague amalgam of "female complaints"—Waring's indictment of the technological and social obstacles to women's physical and environmental health found its way, by direct and indirect citation, into a range of publications for widely varying audiences across North America, male and female, lay and professional, through the 1870s. The *Canada Lancet: A Monthly Journal of Medical and Surgical Science*, published out of Toronto, for example, reprinted Storer's editorial, including the unattributed quotations from Waring's advertisement, under the title "Domestic Latrinæ" (1870).

In 1872, Waring's critique would centrally inform a chapter of the US *Report of the Commissioner of Agriculture for the Year 1871*, issued by the Government Printing Office in Washington, DC. Included as one of twenty-two special reports that were "deemed most complete in matter and most suggestive in [their] teaching" from the preceding year, "Moule's Earth-Closet System" (prepared, according to the report's editor, with Waring's expert assistance) begins in familiar rhetorical territory. "At the bottom of the garden, or at some other inconvenient distance, stands—a temple of defame—the common privy," intones the report. "This necessary resort, even of delicate women, whose condition should command our greatest care, is approached by a path that is often blocked up with snow, deep with mud, or overhung with dripping trees, or overgrown with wet grass." These physical discomforts (always wet and frequently dripping) are accompanied, as in Waring's own writings, by the social evil of unwanted male observation: "If the house is in a village, this walk is probably so exposed to the public gaze that women are often tempted to postpone their visits until nightfall. This and the neglect that comes of the dislike to encounter cold and wet are fertile causes of the ill-health for which country women are especially noted." The government report (which one suspects may have been drafted not merely with Waring's aid but also through his dictation) expressly includes women in its call for political activism. "The proposed reform" of these conditions, the report insists—which it hoped to see initiated locally and then applied nationwide, privy by privy, either through replacement or renovation—"should secure the best efforts of all sensible men and women" (US Department of Agriculture 1872, 123, 466, 467).[9]

Even as Waring's views gained cultural traction, however, one can discern the beginning of a gradual retreat—at times by Waring himself—from the most radical implications of his call for women's political and environmental activism. For instance, the advertisements that Waring wrote for the Earth-Closet Company multiply the "horrors" of the privy, but make no explicit mention of women's rights, indicating that at least in this context Waring preferred to target women as customers rather than as political agents. Both the earth-closet ads and the *Report of the Commissioner of Agriculture*, moreover, direct their closest attention to "country women," hinting at a class bias that reappears in the report, in particular, with condescending force.[10] "It is true that our middle and lower classes are better fed"—as well as better clothed, housed, and educated—"than those of most other countries," the report allows. "But, for all this, they not only lack some of the most important comforts and decencies of life; they do not even know that they lack them," especially in regard to environmental health. "With bodies that are susceptible to the poisonous influences of putrefying filth, with their health more or less constantly subject to these influences, . . . they live on, indifferent to, if not ignorant of, the dangers and discomforts that surround them." Americans of these classes "are busy in accumulating the means for more luxury while they remain blind to improvements which, costing comparatively little, would prolong their lives, secure exemption from disease, and make their homes much more fit abodes for an intelligent and prosperous people" (US Department of Agriculture 1872, 466). To the extent that these remarks reflect Waring's views, they would seem to belittle, not empower, the majority of his potential female readers.

We can trace this gradual retreat most clearly in the writings of other prominent commentators who took up Waring's call for privy reform over the remainder of the decade, usually reshaping it according to their own interests. Instead of emphasizing a woman's right to personal and environmental health, say, many commentators tended to concentrate their rhetorical energy on the commode's filth. When William Goodell, clinical lecturer on the diseases of women and children at the University of Pennsylvania, made women and privies the focus of his often-reprinted 1873 lecture, "On the Relation Which Faulty Closet-Accommodations Bear to the Diseases of Women," his account of the physical barriers to women's health seems calculated to outdo Waring's own in expressive power. After quoting briefly from the same Earth-Closet Company advertisement that had caught Boston gynecologist Storer's attention, Goodell (1873, 738) pulls out every rhetorical stop in denouncing the state of women's private "accommodations":

The very name of *privy* is a misnomer. How seldom is the building hidden by clumps of evergreens or masked by any other disguise than that of a euphemism! How often is it not at an embarrassing distance from the house,—at the end of a long trail, or, at best, of a long and ill-kept path, which frequently runs parallel with a street or a road! How rarely is it ever provided with any other kind of window than round or crescent-shaped holes rudely cut out of the door! How commonly are the cracks dehiscent! The door itself is often without a bolt, often hanging by one hinge,—sometimes wholly unhung. Through the openings in the seat acrid blasts of wind sweep up as if impelled by some malignant demon. Now, add to all this the sickening stench, the conspicuous heaps of filth, the swarms of unclean flies, and confess that, despite the temerity of the description, the picture is not overdrawn.[11]

Breathlessly improvising on Waring's original snapshot of the remote and inhospitable privy, Goodell's virtuoso portrayal offers up fresh details (rudely cut windows! dehiscent cracks! acrid winds!) with a melodramatic flourish of exclamation points. (At a later point in the lecture, he describes using such a privy as akin to being "buffeted of Satan.") What is decidedly not "overdrawn" in Goodell's account, though, is the depiction of women's specific roles in the eradication of these evils. Although from the text of his lecture it would appear that Goodell included a practical demonstration of the earth closet for his audience ("at my request, this gentleman has kindly consented to exhibit to you the mechanism"), he does not encourage women to take the lead in privy reform (as Waring had urged) by either building their own commodes or campaigning for the replacement of unsanitary privies. Instead Goodell imagines women merely "shrinking," in an appropriate display of a "womanly sense of decorum," from the discomforts of the modern outhouse. He envisions physician-experts like himself playing the leading role in reforming toilet habits in the United States. "As our title, *doctor*, indicates," Goodell (1873, 738–39) explains, "we ought to be the teachers as well as the healers of the community,—the educators and the refiners of those among whom our lot is cast."[12] Under Goodell's call to action, women might benefit from reform, but they will not be called to direct it.

In a similar fashion, when journalist and public health advocate James C. Bayles, an open admirer of Waring's *Earth-Closets and Earth Sewage* pamphlet, turns to the US privy in his popular 1878 volume *House Drainage and Water Service in Cities, Villages, and Rural Neighborhoods*—the first standard work in the United States on the mechanics of hygiene—he reserves his most evocative language for the description of waste, not political action or environmental rights. "I know of nothing more disgusting to sight and smell, more nauseating to the stomach or more dangerous to health," fumes Bayles, "than a typical country privy, with its quivering, reeking stalagmite of excrement under each seat, resting on a bed of filth indescribable." In the face of such filth, Bayles figures women not as environmental activists but rather as passive victims, incapable of breaking the debilitating cycle of constipation and irregularity. "It is not an uncommon thing for women in the country to allow themselves to become so constipated that days and sometimes weeks will pass between stools," Bayles notes. "It seems to be a tendency of the sex which easily assumes the form of a habit." Much as Goodell called on physicians to lead privy reform, Bayles (1878, 267, 268, 269) vests responsibility in the (presumably male) head of household: "Until we provide our families with better facilities than are now commonly enjoyed by them, the important duty of daily evacuation of the bowels will be neglected in wet and cold weather by all who can find any excuse for so doing."[13]

Popularizers of Waring's attack on the privy further diluted his call for women's activism by emphasizing individual morality over group rights, comfort over equality, and jingoistic pride over civic participation. "Wage a successful crusade against these affronts to health and to decency," Goodell (1873, 738, 739) declared at the conclusion to his lecture, "and you give better bone, better nerve, and better muscle to the state, and better morals to the people." In the lead editorial of the same issue of the *Philadelphia Medical Times* that first published Goodell's address, for example, the journal's editor praised the doctor's emphasis on morality by contrasting appropriately modest US women with their shameless European counterparts. Recalling, on a recent trip to Europe, being an abashed witness to "a lady who was worshipping with unclosed portals in the temple of Cloacina"—the double entendre of the overwrought metaphor suggestively linking the open doors ("unclosed portals") of the privy to the exposed workings of the woman's own body—the editor reminds his *Medical Times* readers that "such absence of shame goes hand-in-glove with license of language and of morals." Without better privies, the editorial (1873, 747) not so subtly insinuates, US women are in danger of turning out just like (gasp) the amoral French.

In his own writings during this period, Waring nonetheless continued to identify women as an integral part not only of his intended audience but also of the solutions to the problems of home and town waste management. In fall 1875, for example, in the first installment of a lengthy three-part series on sanitary drainage that Waring contributed to the *Atlantic Monthly*, he made a point of including women in his definition of the "general public" he wished to enlist in the reparation of sanitary neglect. "Happily men, and women too," Waring (1875, 339) writes, "are fast coming to realize the fact that humanity is responsible for much of its own sickness and premature death, and it is no longer necessary to offer an apology for presenting to public consideration a subject in which, more than in any other, . . . the general public is vitally interested." Although in the rest of the series Waring relies on the generic "he" to identify the householder whose sanitary acumen he hopes to train, this unusual opening gesture ("and women too") suggests Waring felt there was not only no need to apologize for introducing indelicate subjects to his readers but also no need to apologize for addressing men and women as equal partners in the eradication of waste-borne dangers. Two years later, in another series of popular articles, this time for *Scribner's Monthly*, Waring would more specifically identify women's civic (rather than exclusively domestic) role in cleaning up US waste. "At the outset it is to be said that the organization and control of the village society is especially women's work," Waring 1877, 98) notes. "It requires the sort of systematized attention to detail, especially in the constantly recurring duty of 'cleaning up,' that grows more naturally out of the habit of good housekeeping than out of any occupation to which man is accustomed."[14]

At the same time, Waring's reliance here on a conception of women's so-called natural habits of "good housekeeping" threatens to diminish, or at least compartmentalize, women's civic responsibilities by making the source of women's interest in public issues seem biological as opposed to political. Indeed, the absence of a return to the explicit rights rhetoric of *Earth-Closets and Earth Sewage* in Waring's popular essays of the 1870s implies that Waring, too, was guilty of slowly pruning away the political language that erupted with such sudden force in his second pamphlet. The fact that no other commentators take up this part of Waring's argument during these years, preferring, as we have seen, to expend their rhetorical energy on increasingly graphic descriptions of privies and excrement, further marks the degree to which Waring's identification of a woman's right may have largely fallen on deaf ears.

Only once during the 1870s did another writer cite the paragraph from Waring's pamphlet in which that phrase appears. In the March 1879 issue of the *Sanitarian*, the unofficial journal of the US sanitary science and public health movement, Storer—whose 1870 editorial in the *Journal of the Gynæcological Society of Boston* had first brought Waring's writings on women and privies to the attention of the medical establishment—reprinted multiple excerpts from Waring's work in his essay "The New 'Protective' Principle in Public Sanitation." He not only reprised his original citations from the advertisement Waring wrote for the Earth-Closet Company ("the quotation which was then presented will bear repetition," he insisted, "even after this lapse of time"), Storer (1879, 104) then quoted freshly and extensively from *Earth-Closets and Earth Sewage*, including Waring's paragraph on a woman's right.[15] Rather than revive interest in

Waring's claim, however, Storer's essay merely archives the final print appearance of this provocative formulation, which would receive no more attention over the next decade. And thus an era that began by promising US women, in Waring's words, "unspeakable relief," seems at its close unable to speak forcefully about the potential relationship between women's rights, civic equality, and environmental health, focusing instead on safety, comfort, and privacy.

The Sanitary Millennium

By 1880, commentary on women and privies had become predictable, even routine. Writers still called on the same tropes of degradation and exposure popularized by Waring a decade earlier, but rather than trying to outdo their predecessors in detail or disgust, new writers tended to compress their descriptions in deference to a readership already familiar with the contours of the problem. Phrases like "as every physician knows" began to substitute for more minute particulars.[16] As sanitary science itself became a more accepted and familiar field of investigation, commentators also began to ask whether they should restrict themselves to scientifically measurable data as opposed to intrinsically vaguer forms of sociocultural analysis. After quickly anatomizing (in appropriately staccato sentences) the physical dangers of remote farm privies for women in his contribution to the *Fourth Annual Report of the Pennsylvania Board of Agriculture*, for example—"On farms, it is a common thing for a privy to be in an exposed position. The members of the family, particularly women, shrink from exposing themselves by passing to and from it. It sometimes stands utterly exposed. The grass bordering a narrow path is wet with dew, or the path itself is muddy, or full of snow"—Ellwood Harvey turns a brief eye to the consequences of those dangers for "decency, . . . manners and . . . morals" before suddenly breaking off. "But it is not strictly within the province of my subject," Harvey (1881, 137) demurs, "to consider the matter from any other than a sanitary point of view."

Thus when Harriette Merrick Plunkett, one of the female pioneers of the early sanitary movement, published *Women, Plumbers, and Doctors; or, Household Sanitation* in 1885, a reader eager for a fresh examination of a woman's right to environmental health might have thought that day had finally arrived. But she would have been wrong. Plunkett's volume, although written "to arouse the interest and practical efforts of a new class"—women—in sanitary science, imagines only a limited political scope for those efforts. Plunkett's ideal reader is the "tender wife and loving mother" eager to do "her whole duty in defending the health of her household." Although Plunkett (1885, 8, 10) hopes this reader will "rise above the beaten paths of cookery and needlework to some purpose," that purpose will be found only inside the home, whose "divinely appointed mission" it is each woman's duty to guide. In the end, Plunkett's reader is beholden to the same surveillant regime Waring prescribed for the housewife of *How to Drain a House*—also published in 1885—whose job it is to watch her servants as scrupulously as she watches her house, her family, and herself for signs of filth. "Eternal vigilance," Plunkett declares, "is the price of everything worth the having or the keeping" (ibid., 43). If anything, Plunkett makes the stakes even more personal. For while "there is nothing in hygiene" that Plunkett's reader cannot comprehend, "too often does she realize this and [only] begin to study it, when, too late, she stands beside the still form of some precious one, slain by some one of those preventable diseases that, in the coming sanitary millennium, will be reckoned akin to murders" (ibid., 10). If the price of not studying is the death of your child, how dare a mother not accept as her purpose a life of perpetual domestic vigilance?

In this climate of retrenchment, it is then perhaps not surprising that when Waring reprised in *How to Drain a House* the influential section from *Earth-Closets and Earth Sewage* in which he first attacked the problem of women and privies—the only time he ever reprinted those passages—he quietly left out the paragraph articulating a woman's right to intervene not just in technological remediation or domestic management but also in the sociocultural habits and prejudices that helped sabotage US women's health in the late nineteenth century.[17] Women readers of *How to Drain a House* would thus have little inkling of the radical potential initially breathed into that small yet provocative pamphlet, whose call for reform had been treated with indifference for fifteen years.

This is not simply to condemn Waring for abandoning a specific political stance. Rather, I have tried to show how Waring's impassioned critique of women's subordination led him to propose a right so significant for women's environmental health that he felt it was more important than the fight for suffrage or jobs; how Waring's critique caught the imagination of multiple commentators, who nonetheless ignored his call for that particular right even while embracing his other sanitary reforms; and how we might therefore understand Waring's gradual retreat from the broader implications of that call when no one else seemed to think it was worth talking about—not even writers whose primary goal was to reshape Waring's core ideas for women readers in the United States.

The powerful late-century fantasy of a sanitary millennium—a much-anticipated arrival of a modern era of cleanliness and hygiene—likely also contributed to the failure of any broader theory of women's extra-domestic environmental rights to flourish during this era. As described succinctly by J. R. Black (1872, v) in his hygienic decalogue *The Ten Laws of Health; or, How Disease Is Produced and Can Be Prevented*, the sanitary millennium was thought to herald a time when "all uncertainty in reference to the preventability of nearly every form of disease must disappear, and the laws pertaining to health be capable of definite settlement; so that men and women, by living accordantly with them, may live as they ought to live, free from disease, and die as they ought to die, from old age, and not by the violent and unnatural process of disease."[18] Historian Nancy Tomes notes that this emphasis on certainty and preventability could evoke strong feelings of guilt in US women, as when Plunkett herself insinuates that any lapse of attention might be tantamount to murder. Since domestic sanitation was so tightly focused on the interior of the home—the space in which US women were thought to have the greatest natural interest and over which they could exert the most control—it is not surprising that less attention might gradually be paid, either by commentators or practitioners, to the collateral (and not strictly hygienic) degradations of the public realm, as epitomized, for example, in the "public gaze" on the privy path that Waring felt invariably compromised a woman's right to maintain the health of her body. Although Tomes (1998, 66) also observes that "this sense of responsibility for guarding the home ultimately impelled" some middle-class women in the United States "to forsake its limited domain," leading them (like Plunkett herself, the spouse of one of the Massachusetts State Board of Health founders) to pursue extradomestic professional careers as public health activists or sanitary writers, the majority of women were directed ever more anxiously—sometimes even, as we have seen, by female public health advocates—back into the home.[19]

Waring's call for a woman's right to personal and environmental health may additionally have been undermined, ironically, by the success of his own agitation for more sanitary privies. As more and more Americans replaced their backyard "temples of defame" with indoor water closets—increasingly the preferred choice over Waring's pet dry-earth system—homeowners now had to contend with an often-bewildering array of pipes, drains, and traps, usually hidden out of sight, beneath the floor and behind the walls. Not only did this create a new mechanical infrastructure for women to supervise—and keep clean—it heightened anxiety about "sewer gas," the germ-carrying vapors from inadequately flushed waste widely assumed to be the cause of most airborne diseases. Waring (1885, 63), whose success in the 1870s developing large-scale town sewerage systems made him a national expert on the proper disposal of waste, railed with special vehemence against the "offensive exhalations" that could seep, undetected, from even the cleanest-looking drains. Hence, in what May N. Stone (1979, 283) has called the "plumbing paradox," the gradual introduction of indoor toilets solved the problem of the exposed privy at the same time that it exacerbated the new interior structures of self-surveillance (and guilt) to which middle- and upper-class US housewives in many cases now found themselves chained.[20]

It would only be during Waring's last and most celebrated civic appointment—his term as the street-cleaning commissioner of New York City beginning in 1895—that his early instinct for anticipating and endorsing women's contributions to the environmental health movement would finally bear public fruit. As Suellen Hoy (1995, 71, 79, 81) notes, Waring possessed a tremendous "faith in the willingness of individuals and communities to engage in sanitary action," and "campaigned tirelessly to involve groups of citizens—especially plumbers, women, and school children—in the work of city cleansing." Thus, when Waring needed public support for his unusual but effective "White Wings" street-cleaning brigade and innovative source-separation recycling program, he made a point of meeting regularly with civic groups, particularly women's organizations. And after Waring was replaced as commissioner by the return of a corrupt Tammany administration in 1898, it was new civic and community organizations led by women, and emboldened by his success, that took the lead in carrying his methods into other cities and towns.[21]

Whether Waring registered the irony of these affiliations and outcomes, given the retreat outlined above, we cannot say. What we can say, however, is that Waring's New York success also places another of his early beliefs in fresh relief—namely, that radical environmental change requires both technological and sociocultural intervention. After all, the "White Wings" kept New York's streets clean not because they had better brooms but rather (in part) because Waring's decision to outfit them in white duck uniforms—a seemingly absurd costume for street cleaners—boosted morale, improved efficiency, and perhaps most significant, reshaped public perception of the importance of the position and its duties.[22] This is in many ways precisely the intangible combination of tactics already implicit in *Earth-Closets and Earth Sewage*'s approach to solving the problem of women's privies: part technology (make them cleaner) and part psychology (redirect the public gaze)—a powerful strategy that women environmental activists in the United States, both despite and because of Waring, would continue to adopt in the decades to come.

Notes

1. For an account of the ironic increase in women's domestic labor produced by late nineteenth-century "improvements" in household technology, see Schwartz 1983, especially 40–68.

2. For works by the few historians who discuss Waring's writing on women and privies, see Handlin 1979, 459–60; Strasser 1982, 94–96. For an excellent overview of midcentury US plumbing, including a discussion of women and privies, see Ogle 1996, 61–92, especially 71–74. For a general look at the social dimensions of late nineteenth-century US sanitary reform (albeit one that does not specifically explore gender), and Waring's impact on urban planning and public health, see Peterson 1979.

3. For more on Waring's early training in scientific agriculture, see Melosi 1977, 6–7. For a concise overview of Waring's career, from his apprenticeship with Mapes to his posthumous apotheosis as the "Apostle of Cleanliness," see Peterson 1995.

4. The 1854 edition of *The Elements of Agriculture* includes an advertisement for a sequel by Waring, *The Earthworker; or, Book of Husbandry*, which in other venues D. Appleton also advertised as "in press," but no such edition ever appeared. It is possible that this was the book Waring intended to write when he finally published *The Handy Book of Husbandry* in 1870, later revised and reissued as *Book of the Farm* (Philadelphia: Porter and Coates, 1877).

5. On Waring's entry for the design competition, see Rybczynski 1999, 176. For an account of Waring's drainage of Central Park and the continued carrying capacity of some of his original pipes, see Biebighauser 2007, 5–6. According to Thomas Biebighauser, Waring is also responsible for creating such external water features in Central Park as the Fifty-Ninth Street Pond.

6. The Beechers devoted chapter 35, "Earth-Closets," to their discussion of Waring's dry-earth system. See Beecher and Stowe 1869, 403–418.

7. In the second edition of this pamphlet, six months after the first printing (and possibly timed to take advantage of the anticipated publicity surrounding the publication of the Beechers' *American Woman's Home* that spring), Waring (1869, vii–viii) introduced into his new preface a series of graphic examples of the restorative power of dry earth to heal wounds, including those of three men who had suffered the near amputation of various appendages and a woman who had "an entire breast . . . removed for cancer." These examples, which only indirectly promoted the earth-closet system itself, seem designed to ratchet up the emotional valence of Waring's broader appeal for the adoption of the earth closet and may suggest an intentional targeting of women readers, whom Waring could assume, now that he had the Beechers' endorsement, might be more likely to buy his publications.

8. I have found copies of the advertisement, for example, on the front page of Connecticut's *Meriden Daily Republican* in consecutive issues running from late December 1869 through February 1870. These ads not only list the company's State Street headquarters in Hartford as well as direct interested customers to additional salesrooms in New York City, Philadelphia, and Boston, they also list as an agent for Connecticut's Fairfield, New Haven, Middlesex, and New London counties one James Waring—likely George's youngest brother. As the Earth-Closet Company had ten different local offices from New Orleans to Albany along with over fifty agencies, it seems likely that the company would have placed this ad (or ads like it) in other local papers during this winter as well.

9. Although the volume editor merely notes that Waring's aid, as the chief promulgator of Moule's theories in the United States, "has been invoked in the preparation of this paper" (US Department of Agriculture 1872, 465), in many places the report recycles material directly from Waring's earth-closet pamphlets. "Moule" is the Reverend Henry Moule, the mid-nineteenth-century British clergyman widely credited with being the first to elaborate a plan for the earth-closet system and whose early designs heavily influenced Waring's own. See, for example, Waring 1870, 9.

10. The advertisements for the Earth-Closet Company always address themselves thus: "TO COUNTRY WOMEN. The following considerations concerning the DRY-EARTH SYSTEM are respectfully submitted as worthy of their thoughtful attention."

11. Goodell's flair for the dramatic is apparent from the opening sentence of the lecture, in which he invokes Edmund Burke on the sublime. "'The sublime,' writes the great Burke, 'is an idea belonging to self-preservation.' Emboldened by this definition, I shall offer no apology for addressing you this morning upon an unsavory subject" (Goodell 1873, 737).

12. Goodell's lecture was not only reprinted frequently through the decade, both in North America and Europe, but a slightly revised version also appeared in the monumental volume by Goodell (1879) titled *Lessons in Gynecology*, the leading text in the field in the late nineteenth century.

13. In the same chapter, Bayles (1878, 274) recommends Waring's *Earth-Closets and Earth Sewage* to his readers: "If a copy of his pamphlet . . . were placed in the hands of every country physician, I am satisfied that great and important benefits would result in drawing the attention of the profession to many things concerning which they are, generally speaking, either ignorant or indifferent."

14. When Waring (1876, 10) published the three *Atlantic Monthly* essays in book form the following year, he accentuated his inclusion of women in the opening invocation of his audience by setting the phrase "and women too" between dashes ("Happily, men,—and women too,—are fast coming to realize . . . "). He also clarified in a new preface that he was writing for "the average citizen and householder" (ibid., v).

15. Although Storer (1879, 106) does not comment on Waring's claims for a woman's right, he does conclude his essay by recommending an innovation he has long thought worthwhile: the use of toilet seat covers for additional "safety

and comfort"—a particular benefit for women, who unlike men sit down when urinating as well as defecating.

16. See, for example, Gerhard 1882, 153. That William Paul Gerhard—a German immigrant who became Waring's chief assistant in 1881—commonly used this phrase, suggests that Waring himself tacitly approved the shift in emphasis. Gerhard's *House Drainage* would become one of the standard late-century texts on home waste management, appearing in seven editions by 1898.

17. As suggested above, as best as I can tell, even when excerpts from Waring's passages on women and privies from *Earth-Closets and Earth Sewage* were reprinted by other writers (with the exception of Storer's reprinting in the March 1879 issue of the *Sanitarian*), they never include the paragraph on a woman's right. For example, the excision of the woman's right paragraph in an article for the *Cincinnati Medical News* by B. K. Johnson (1879) is all the more revealing since Johnson clearly borrowed his Waring passages from Storer—who of course *did* include that passage. (That Storer is Johnson's source is evident from the end of the selection, which includes Storer's musings on the need for toilet seat covers.).

18. Black himself does not use the phrase "sanitary millennium" in his treatise, but an unnamed reviewer highlighted precisely this paragraph as an apt definition of the concept. See *Boston Medical and Surgical Journal* 10, no. 17 (October 21, 1872), 289.

19. On Plunkett's background, see Logan 1912, 902.

20. Between 1870 and 1880, Waring supervised the installation of sewers in cities from Saratoga Springs, New York, to Memphis, Tennessee. His controversial advocacy for "separate" conveyance systems (such as the system he installed in Memphis in 1880) brought him such widespread national attention that when the White House was suspected of having a sewer-gas problem, Waring was called to Washington, DC, to conduct an inspection. See Melosi 1982, 56–78; Tomes 1998, 72–75.

21. For more on Waring's relationship to Progressive Era civic reform, see Burnstein 2006, 32–54.

22. See, for example, Skolnik 1968, 362.

References

Bayles, James C. 1878. *House Drainage and Water Service in Cities, Villages, and Rural Neighborhoods*. New York: David Williams.

Beecher, Catherine E., and Harrier Beecher Stowe. 1869. *The American Woman's Home: Or, Principles of Domestic Science*. New York: J. B. Ford.

Biebighauser, Thomas R. 2007. *Wetland Drainage, Restoration, and Repair*. Lexington: University of Kentucky Press.

Black, J. R. 1872. *The Ten Laws of Health; or, How Disease Is Produced and Can Be Prevented*. Philadelphia: J. B. Lippincott.

Book Notices. 1870. Review of *Earth Closets and Earth Sewage*, by George E. Waring Jr. *Little Corporal* 10, no. 5 (May): 77.

Burnstein, Daniel Eli. 2006. *Next to Godliness: Confronting Dirt and Despair in Progressive Era New York City.* Urbana: University of Illinois Press.

Domestic Latrinæ. 1870. *Canada Lancet: A Monthly Journal of Medical and Surgical Science* 3, no. 3 (November): 90–91.

Earth-Closet Company. 1869. Advertisement. *Meriden Daily Republican* 27 (December): 1.

Editorial. 1873. Privies and Morals. *Philadelphia Medical Times* 3 (47): 746–747.

Gerhard, William Paul. 1882. *House Drainage and Sanitary Plumbing.* New York: D. Van Nostrand.

Goodell, William. 1873. On the Relation Which Faulty Closet-Accommodations Bear to the Diseases of Women. *Philadelphia Medical Times* 3, no. 47 (August 23): 737–739.

Goodell, William. 1879. *Lessons in Gynecology.* Philadelphia: Brinton.

Handlin, David P. 1979. *The American Home: Architecture and Society, 1815–1915.* Boston: Little, Brown.

Harvey, Ellwood. 1881. Sanitary Arrangement of Farm Buildings. In *Agriculture of Pennsylvania: Fourth Annual Report of the Pennsylvania Board of Agriculture for the Year 1880*, 134–139. Harrisburg, PA: Lane S. Hart.

Hoy, Suellen. 1995. *Chasing Dirt: The American Pursuit of Cleanliness.* New York: Oxford University Press.

Johnson, B. K. 1879. Water Closet Conveniences. *Cincinnati Medical News* 8, no. 9 (September): 619–621.

Logan, Mrs. John A. 1912. Harriette M. Plunkett. *The Part Taken by Women in American History*, 902–903. Wilmington, DE: Perry-Nalle.

Melosi, Martin V. 1982. *Garbage in the Cities: Refuse, Reform, and the Environment, 1880–1980.* College Station: Texas A&M University Press.

Melosi, Martin V. 1977. *Pragmatic Environmentalist: Sanitary Engineer George E. Waring, Jr.* Essays in Public Works History, No. 4. Washington, DC: Public Works Historical Society.

Ogle, Maureen. 1996. *All the Modern Conveniences: American Household Plumbing, 1840–1890.* Baltimore: Johns Hopkins University Press.

Peterson, Jon A. 1979. The Impact of Sanitary Reform upon American Urban Planning, 1840–1890. *Journal of Social History* 13, no. 1 (Fall): 83–103.

Peterson, Jon A. 1995. Waring, George Edwin, Jr. In *Pioneers of American Landscape Design 2: An Annotated Bibliography*, ed. Charles A. Birnbaum and Julie K. Fix, 155–164. Washington, DC: US Department of the Interior, National Park Service, Cultural Resources, Heritage Preservation Services, Historic Landscape Initiative.

Plunkett, Mrs. H. M. 1885. *Women, Plumbers, and Doctors; or, Household Sanitation.* New York: D. Appleton.

Rybczynski, Witold. 1999. *A Clearing in the Distance: Frederick Law Olmsted and America in the Nineteenth Century*. New York: Scribner.

Schwartz, Ruth Cowan. 1983. *More Work for Mother: The Ironies of Household Technology from the Open Hearth to the Microwave*. New York: Basic Books.

Skolnik, Richard. 1968. George Edwin Waring, Jr. *New-York Historical Society Quarterly* 42, no. 4 (October): 354–378.

Stone, May N. 1979. The Plumbing Paradox: American Attitudes toward Late Nineteenth-Century Domestic Sanitary Arrangements. *Winterthur Portfolio* 14, no. 3 (Fall): 283–309.

Storer, Horatio R. 1870. Editorial. *Journal of the Gynecological Society of Boston* 3, no. 4 (October): 270–272.

Storer, Horatio R. 1879. The New 'Protective' Principle in Public Sanitation. *Sanitarian* 7, no. 72 (March): 99–106.

Strasser, Susan. 1982. *Never Done: A History of American Housework*. New York: Pantheon.

Tomes, Nancy. 1998. *The Gospel of Germs: Men, Women, and the Microbe in American Life*. Cambridge, MA: Harvard University Press.

US Department of Agriculture. 1872. Moule's Earth-Closet System. In *Report of the Commissioner of Agriculture for the Year 1871*, ed. Frederick Watts, 465–496. Washington, DC: Government Printing Office.

Waring, George E., Jr. 1854. *The Elements of Agriculture: A Book for Young Farmers, with Questions Prepared for the Use of Schools*. New York: D. Appleton.

Waring, George E., Jr. 1867. *Draining for Profit, and Draining for Health*. New York: Orange Judd.

Waring, George E., Jr. 1868. *Earth-Closets: How to Make Them and How to Use Them*. New York: Tribune Association.

Waring, George E., Jr. 1869. *Earth-Closets: How to Make Them and How to Use Them*. 2nd ed. New York: Tribune Association.

Waring, George E., Jr. 1870. *Earth-Closets and Earth Sewage*. New York: Tribune Association.

Waring, George E., Jr. 1875. The Sanitary Drainage of Houses and Towns I. *Atlantic Monthly* 36, no. 215 (September): 339–356.

Waring, George E., Jr. 1876. *The Sanitary Drainage of Houses and Towns*. New York: Hurd and Houghton.

Waring, George E., Jr. 1877. Village Improvement Associations. *Scribner's Monthly* 14, no. 1 (May): 97–107.

Waring, George E., Jr. 1885. *How to Drain a House: Practical Information for Householders*. New York: D. Van Nostrand.

3

Enviroblogging: Clearing Green Space in a Virtual World

Stephanie Foote

I

In April 2010, *Discover* magazine ran a story called "Museum-Worthy Garbage: The Art of Overconsumption" showcasing a group of artists who call attention to "the scale of our collective daily consumption and waste" by "gathering up the trash that washes up on beaches and digging through their own garbage cans" in search of material to transform into art. Featuring photographs of work ranging from pointillist-inspired "paintings" made of scavenged aluminum cans to wreaths made of gaudy bits of discarded plastic, the article also includes brief statements from the artists about their choice to use postconsumer waste as a creative medium. Although some of the artists discussed the aesthetic challenges of working with garbage, almost all of them believed that their choice had an important political dimension. Working with garbage, they agreed, makes visible the wastefulness of US culture. But it's also a metaphor for what they believe to be the precarious, even endangered status of art itself as a way to generate crucial public conversation about social issues. Transforming what is valueless into something beyond price is a way to critique a commodity culture that generates more and more waste as well as draw attention to how those transformations inspire new kinds of conversations about what counts in late capitalism as a wasted object—and in the case of art itself, what counts as a wasteful or nonproductive practice.

Few environmental policymakers, of course, would argue that using postconsumer garbage in art installations has any serious impact on large-scale industrial issues of garbage. Indeed, few artists would make such a claim, and yet their work should be taken seriously not just as a concrete example of the increasing public visibility of environmental issues but also as a sign of the range of idioms and representational practices people are using to generate public debate. It is thus perhaps as interesting that the media spotlight artists working in garbage—or fashion designers who use trash to design clothes, or public installations that draw attention to it like the life-size "trash people" of HA Schult (http://www.haschult.de/trash.html)—as it is that artists are working in the medium of trash. It tells us (in the rhetoric of the very consumption that artists are working to criticize) that there is now a broad enough audience for environmental issues, especially those that constellate material waste and wasteful practices, to make such a story worth covering. And it tells us that the everyday practices of consumption and waste are now part of a dialogue about issues that necessarily implicate individual social actors in a larger understanding of global patterns of consumption, circulation, and waste.

Up-cycling postconsumer garbage into art meant to inspire public dialogue is an example that indicates an interest in public conversation about waste, but in some sense, it is also an instance of the limits of such conversations. What, after all, are different people talking about when they talk about garbage? Are they telling the same story about it? Do they even agree on what it is? Not everyone who is concerned about garbage is an artist, not every act of weighing the impact of an individual piece of garbage will be displayed for other people's consideration, and not every kind of waste can be so easily translated into a politically inspired artistic commentary. And yet although people who throw their recyclables in the correct bin probably do not consider themselves as making an artistic statement, they nevertheless exist in the same horizon of environmental awareness as artists, who most probably understand themselves as recycling. More critically, we might even say that those subject positions—recycler and artist—share a commitment to an aesthetic of self-improvement and civic cultivation—an aesthetic that is derived from one's sense of individual agency in the world as well as anxiously dependent on one's sense of interconnectedness with a world of anonymous others.

We might understand the larger turn toward individual recycling along with the use of garbage and waste as public art as a convergence in the project of what Michel Foucault (1988) called "the care of the self," and the social technologies that make that project widely available to a certain class and kind of individual. An artist working in garbage and an individual social actor for whom recycling has become not only a private virtue but also a public commitment might then share a number of attachments to what are sometimes called "green politics," although they will be interpellated by those politics quite differently. Each is simultaneously positioned as an audience for various green media, as part of a public that can be addressed by officials who prioritize environmental issues, as part of a consumer demographic that might be considered susceptible to environmental pleas, and as an individual who believes that her choices will have some impact on larger social and cultural issues. But each will be engaged, though at different levels of intensity, in a project of self-cultivation that is by necessity driven by the recognition that individual choices about material objects—how to buy them, consume them, and get rid of them as well as how to talk about them, and with whom—are part of an information economy in which they are connected in countless ways to people and corporations they cannot ever really know.

The constellation of technologies of the self and communication technologies is central to this chapter. I will examine the medium of green blogs and Web communities in order to look at how subjects negotiate a sense of individual agency and participation in larger political and social structures when they join online communities. But I want to pause here for a moment to discuss the issue of interpellation, or how subjects come to consent to how they are identified by larger organizations and structures. As political and cultural theorists have argued, how social actors are interpellated as members of a public as well as an audience or demographic is not the same as how they understand their own political affiliations and commitments; in some sense, the social imaginary of what we might call everyday environmental awareness is composed of a constant struggle between how social actors are addressed by various political and social groupings, and how they choose to inhabit subjectivities and create communities based on those environmental commitments. And as scholars like Michael Maniates (2002), Thomas Princen (2002), and Andrew Szasz (2007) have contended, organizing personal practices and social and political commitments around consumption is a fraught, yet characteristic mode to express environmental commitments at this moment in the United States.[1] To take one example, deliberately consuming less appeals to people searching for ways to live a more environmentally friendly lifestyle, and provides a concrete way for individuals to feel that they can make a measurable difference in a vaguely defined but nonetheless intensely felt desire to live sustainably by controlling wasteful personal practices.

Although environmental writers and thinkers have argued against the idea that making small changes in one's lifestyle, patterns, or rate of consumption can effect long-term environmental change, adopting a more green or "ecofriendly" set of lifestyle practices has become has become more popular as a kind of political statement precisely because it has become more available to consumers as a practice of everyday life.[2]

Consumers concerned about everything from wasting energy, to toxic household cleaning solutions, to finding local organic food solutions, to composting their food scraps, can now find answers to their questions in periodicals ranging from the *New York Times* to online news aggregators like Grist.org that specialize in stories about environmental issues. But it is not, as I will discuss, just the fact that more information is available online that has helped consumers believe that lifestyle choices equate with political action. Consumers have found ways to intervene in social media in order to test strategies for the viability of that conversion, and engage in public discussions about the twinned projects of self-cultivation and environmental awareness—projects that often seem to be in tension.

I began by describing how artists have been using waste and garbage as ways to produce public conversation and public discourse about structural problems, and in the rest of this chapter, I will inquire more closely into how blogs about living sustainably by controlling consumption and waste are putting the cultivation of an environmental subjectivity into narrative. Specifically, I focus on a handful of bloggers and Web sites dedicated to confronting large-scale environmental issues by ruminating on individual acts of consumption. I will look at how environmental blogs, which are often focused on consumption and individualized practices of refusals of consumption, reveal the shifting definitions of waste in a broadly defined environmental movement. My aim here is to argue that in the always-expanding world of individual blogs, the connectedness that helps individuals in the United States gather information about personal acts toward living sustainably can also help to model a larger, less individualistic response to an environmental crisis, even though blogs by their very nature tend to think at the individual level as well as in terms of affective and personal responses. Unregulated by the protocols of conventional media, generally open to public commentary from readers they do not know, and sometimes charged by "experts" with circulating mere informational garbage, envirobloggers chart the emergence of increasingly sophisticated thinking around key environmental issues, demonstrating how specific sites of intense environmental anxiety, like waste, are taken up, examined, and shared in new online communities that are not merely reporting on environmental issues but also participating in the evolution of how those issues are disseminated and debated.

As I explore green blogs, I will pay particular attention to how they understand waste in both its material and metaphoric forms, and how they use it to foreground individual consumption as one of the engines of environmental and economically unsustainable practices. But I will also concentrate on something of which bloggers are all too aware: the danger that individual, small-scale choices will take priority over the need for large-scale institutional change. Environmental bloggers are generally committed to the small and local; they champion local foods or local merchants, for example, and almost all the bloggers I've consulted for this chapter (about fifty, though I will discuss only a few) write about their commitments to their little corners of the world even as they recognize the global scale of the environmental crisis. Almost all the blogs ask a similar question: Can changes in individual or household practices make a difference in individual lives, and connect individuals to larger issues and communities? One way to approach this question, they maintain, is to follow and comment on debates in alternative and conventional online and print media about the greening of lifestyles, paying particular attention to debates over the political impact of such choices.

These blogs have much to tell scholars, critics, and activists about how, for instance, corporations respond to perceived consumer demand for greener products, how green practices can be privatized, or even the reach and popularization of environmental commitments that were once perceived to be politically extreme. But we can only do so by looking at how these bloggers make decisions on a day-to-day basis and struggle to reconcile their individual practices with the larger, sometimes-overwhelming scales of change that their lifestyle practices suggest to them. How then, we can ask, is "green space"—a space for green communities and individuals—made and contested in a virtual world? How does that green virtual world map onto or affect the material world in which green ideas are debated and lived? What does individual participation in the blogosphere reveal for our understanding of how people organize around environmental issues, and how we can measure the global effects of local choices? What does the combination of technology and technologies of self-cultivation tell us about the evolution of environmental awareness?

II

From its multiform origins in different modes of posting user-generated content on the Web in the mid-1990s, blogging has emerged as an extraordinarily popular way to publish one's own thoughts in relative privacy as well as, paradoxically, instantiate a virtual community that might first only include friends and family, but that has the potential to include people the blogger has never met (and may never meet) face-to-face.[3] Would-be bloggers can choose to launch their blogs from a variety of servers and can now choose from a multitude of templates. Bloggers can self-publish on sites supported by more conventional news sources, such as the blogs collected in Open Salon, a feature of the popular online daily periodical *Salon.com*. They can launch their blogs for free on a site supported by Google.com or pay for a dedicated service like Typepad. It is now quite easy to launch a blog; free templates are so readily available that no advanced computer knowledge is required, although blogging certainly depends on basic computer literacy and access. While it is difficult to get exact figures on how many blogs there are at a given moment, Technorati.com, the site devoted to tracking the blogosphere, estimates that there are currently millions of blogs, and that every day hundreds of people join the blogging community by beginning their own blogs, or commenting on or just reading one or more blogs.[4] A great many blogs last for only a handful of entries (although they may well stay up on the Web permanently unless their creators go to the trouble of taking them down), but many go on to establish regular postings and comments, and a few blogs are so successful that they lead to more lucrative writing assignments.

It's something of an understatement, then, to say that Weblogs are popular, but the world of blogs is incredibly diverse, making it nearly impossible to predict which ones will gain a broad audience or have mass appeal. Environmental or green blogs are no different; there are now thousands of bloggers focusing exclusively on some aspect of environmental issues. The envirobloggers who concentrate on issues broadly related to the environment cover a great deal of ground, yet most are at least nominally interested in the effects of overconsumption, the lifecycles of the products they buy and discard, and the global meaning of their local choices. Many envirobloggers, though not all, are what would be considered amateur writers. A few are professionals in some area connected to the environmental movement. Some, for example, are builders or engineers looking at emerging environmental practices in their professions. Some live in cities, some live in rural areas, and still others in the suburbs. Many follow the organized green movement as well as read and comment on a range of other blogs (and link back to their own blog), a tiny number have made lucrative book and film deals based on their blogs, some have attained a well-respected presence in the environmental blogosphere, and many more appear to be individuals who sometimes post their affective and critical responses to environmental issues. But those bloggers who are posting about lifestyle changes—whether they post a lot or a little, or whether they are written up in their local papers or interviewed on

National Public Radio—by necessity confront and meditate on the limits and possibilities of how making lifestyle changes can impact large-scale environmental systems.

Most environmental bloggers in the United States are like the vast majority of bloggers (and active members of organized mainstream environmental movements): they are well educated, white, and middle class, with access to the technology that allows them to create and read blogs along with the computer literacy that allows them to post. The bloggers I will discuss here are not unaware of their privilege. Many of them recognize that their criticism of capitalist values and practices is the result of how well they have reaped educational benefits from them—that is, their critiques are enabled by their access to an objectively good and socially powerful set of interpretive and research skills. Thus, while many of the envirobloggers I examined in the process of writing this chapter have made a conscious decision to leave behind the values and often infrastructural amenities of middle-class life by developing alternative methods of energy consumption, for example, they are nearly all uniformly aware that their ability to make this choice is itself a signal of how deeply their class privilege has created them as the sort of people who can express their political will by making such a decision and then publicly reflecting on its ramifications.

Few of the envirobloggers I examined (although some) write exclusively about garbage, but almost all of them discuss household waste and how to keep goods out of the waste stream—reducing or eliminating goods with packaging as well as basic issues of composting, recycling, repurposing, repairs, and so forth. Most, even those keeping more personal or lifestyle blogs, understand their blogs as political, even though many are not activists in the conventional sense of face-to-face local community work. Understandings of the relationship between the personal and political, or the individual and structural, are not uniform; indeed, the location of those boundaries along with their relative porosity and ambiguous demands on social actors are hotly contested on green blogs, and are perpetually rewritten by the medium of blogs themselves, which can be accessed by anyone with Internet capacity.

Most green lifestyle blogs are classically journal like, recording the observations of a single author in a time- and date-stamped format. These blogs usually welcome comments, and as with most blogs, many of the comments appear to come from other like-minded bloggers and people the blogger knows in real life (many of those commenters reveal their relationship to the blogger in the commentary). These blogs are often meditative in nature, describing everyday green choices—frequently those choices that are made in the domestic sphere. These decisions range from what to purchase or make, to how to clean the kitchen sink, to how to determine the carbon footprint of a single-family house. Many are written by people who are attempting to change their entire family's relationship with consumption—some writers are living off the grid (*Little Blog in the Big Woods*) or largely off the grid (*NoImpactMan*); some are involved in sustainable living through organic farming (Sharon, who writes under the name "jewishfarmer" at *Casaubon's Book*), and others are keeping track of food waste (*Wasted Food*). There are also a large number of blogs about zero waste or voluntary simplicity (*Zero Waste Home: Refuse, Refuse, Refuse, Then Reduce, Reuse, and Recycle (and Only in That Order)*, *My Zero Waste*, or *Clean Bin Project: Our Zero Waste, Consumer-Free Year*), and many more about the intersection between frugal living and green lifestyle choices, like *Cheap Like Me: Where Economy and Ecology Meet the Good Life*.[5]

Of the many lifestyle blogs about environmental issues, I concentrate on two of the most widely read—those by Colin Beaven and Vanessa Farquharson—to detail some of the paradoxes involved in trying to create green space in the blogosphere. In some ways, Beaven and Farquharson are typical of envirobloggers in their attention to the details of daily life and attempts to account for the choices they make, and especially the ways that their initial focus on purifying their lives brought them into a perpetually widening set of reflections about what constitutes the "local" world they inhabit. Each began with their households, and inevitably each started to inquire into how commodities mediated their relationships with their neighborhoods, cities, and nations. But they are also different than many envirobloggers. Beaven and Farquharson built a large readership, posted regularly, and have managed to demonstrate the porosity between various kinds of media and expertise. Both of them are professional writers, and each ended up with book deals from their blogs. Beaven is popular on the lecture circuit and participated in a documentary film about his "noimpact" experiment, and Farquharson is a reporter at the Toronto *National Post*. Beaven, who as NoImpactMan set himself the task of reducing his family's carbon footprint to zero, blogged about his year living off the grid living in a New York City apartment with his wife and three-year-old daughter. Farquharson too, participated in a yearlong experiment in reducing the carbon footprint of her life. Their blogs are quite different in tone—Beaven is more prone to earnest preaching; Farquharson is more ironic and tends to make practical suggestions—but together they can help us to see how the technologies of the Web and selfhood come together around an environmentalism that begins with the individual, and by necessity reaches outward.

Beaven's masthead reads,

A guilty liberal finally snaps, swears off plastic, goes organic, becomes a bicycle nut, turns off his power, composts his poop, and while living in New York City, generally turns into a tree-hugging lunatic who tries to save the polar bears and the rest of the planet from environmental catastrophe while dragging his baby daughter and Prada-wearing, Four Seasons–loving wife along for the ride. (http://noimpactman.typepad.com)

The blog is carefully laid out, and now includes a list of public interviews and appearances Beaven will be attending as a guest speaker, various ways to obtain the book or film based on his experiment, and advertisements for events that try to replicate his experiment (you, too, can try to live a week with no impact). The blog's entries are archived by categories, many of which overlap with one another, so that most posts are cross-tagged and referenced. One of Beaven's most regularly tagged categories (tags index blog entries for readers searching by a particular topic) is "Waste Not, Want Not," under which he files entries on the various ways he and his family tried to cut down on waste and consumption. In a blog that is dedicated to chronicling the effort to live a no-impact life, the issue of garbage looms large for Beaven, and he approaches it by trying to get to the root of the issue, which for him is overconsumption.

Much of Beaven's blog tends to make distinctions between himself and his readers—he believes that few people could sustain such a lifestyle for long—at the same time that it challenges individuals to make small changes in their own lives or contribute suggestions in the blog's comments section that others can use. Yet for Beaven, each post is often an opportunity to meditate on how blind he was before the experiment; human beings, he argues, will be happier when they want and have less, live simply, and reject the drive to consume at an unsustainable rate. They must make a personal change within themselves, and then work outward to link that change to their families, neighborhoods, and lawmakers. Although Beaven supports large-scale structural changes that can be initiated when constituents pressure lawmakers, he also prefers to lead by example; the green space he envisions is first in one's mind and then in the world. The titles of two consecutive blog posts reveal the tensions in Beaven's no-impact experiment, for they point out the difficulties of trying to create systematic environmental change as an individual, even as they rely on the individuals who make up the community that blogs both build and address. On September 9, 2010, Beaven posted an entry titled "When Saving the World Threatens to Overwhelm You," and on September 29, 2010, his next post is titled "Come Save the World with Me on 10/10/10."

If Beaven's blog asks readers to change themselves from within even as it assures them that such change will have a broad impact if individuals find a way to somehow publicize that change, it also serves as a test case for the potential and limits of how an individual can create a community of like-minded readers. We might contrast Farquharson's blog—her project is to implement one green innovation every day for a single year—with Beaven's work. Unlike Beaven, who began his blog as part of a multimedia publication deal, and used it to meditate on human nature, global peace, and the enlightenment that follows from giving up wasteful practices, Farquharson deliberately tested the impact of small changes without making claims for their global significance. Farquharson's blog was titled *Green as a Thistle*, and her masthead reads, "Making one change every day to greenify my life (and hopefully not being too smug about it)." Part of her description of her blog's purpose is:

Spend each day, for an entire calendar year, doing one thing that betters the environment. The idea is that everything I do, I keep doing (so if I switch brands, it's a permanent switch; if I turn down my thermostat, I keep it down), so that by day 365, I'll be living as green a lifestyle as it gets. I hope, in the end, this proves that being an environmentalist doesn't necessarily have to require massive change, compromise or Greenpeace levels of dedication—it can be simple, and inspiring. (http://greenasathistle.com)

Green as a Thistle closed after the 365-day project, but reopened in September 2008 with an update on which of Farquharson's green changes she managed to continue, and with news that she had landed a book contract. Her book, *Sleeping Naked Is Green*, was released from a major press in 2009.

Green as a Thistle, while more modest in its aims than Beaven's blog, might arguably have had a more powerful local impact, for rather than simply giving up creature comforts like electricity, as Beaven did (and convinced his somewhat-dubious wife to do as well), Farquharson searched for small local companies making such household objects as soap, shampoo, and furniture, restaurants trafficking with local economies, or larger corporations keeping track of how their products are sourced and delivered. She attempted, that is, to persuade her readers to consider the entire life cycle of objects and commodities, paying special attention to the calculus of embodied energy (how much energy is used in the production, transportation, selling, and disposal of objects). By promoting modest changes as opposed to the emotional exaltation of giving things up, Farquharson pointed out the fact that online communities might be able to connect to and support an emerging local community of environmentalists in a shared material space.

An emphasis on streamlining patterns of consumption also connects envirobloggers to communities other than real ones. Envirobloggers who read and list one another's blogs on their blogrolls (which can direct readers to similar kinds of blogs, indicate political or personal affiliations and approval, or just promote a broad sense of community) deliberately intersect with other virtual communities—in particular, the green-tech online community. Green innovation and technology blogs feature reports on new products or prototypes culled from the international media, and are organized around such keywords as "innovation," "design," and "style." Although they tend to report on products rather than how projects are used, they also tend to focus more on the entire life of a commodity, paying close attention to the energy consumption and green practices of the factories at which various products are made. If the envirolifestyle blogs tend to wonder how to choose within a marketplace of already-made goods—that is, worry at the point of individual consumption—the green-tech blogs tend to have a longer temporal view of consumption as well as a focus on larger structures of research and development. Examples of this sort of blog include ecogeek.org (under "about us," the founders write: "Technology can be a force for evil, or for awesome. Those who shun the tech are just as guilty as those who ignore the environment. There's a safe balance, where the awesome can help nature as much as it helps us have a good time and live easier lives").[6] If the envirolifestyle blogs are influenced by the traditional blog as personal journal, the green-tech blogs are influenced by the vast network of DIY Web culture. This culture tends to be tech oriented as well, and includes such sites as howstuffworks.com (the header of which says simply "it's good to know"), instructables.com ("the world's biggest show and tell"), and makeit.org. The green-tech blogs are interested less in wasteful individual practices and more in how to extend the life cycle of products by reusing them.

Bloggers mostly interested in lifestyles and personal changes read the tech blogs, and the tech bloggers, keenly aware of the growing market for green products and services, often read and comment on lifestyle blogs. There are a fair few number of bloggers who combine the characteristics of each kind of blog: the writer of *Little Blog in the Big Woods*, for example, details his day-to-day life in his sustainable, self-built cottage, yet also describes in great detail the maintenance and development of various technologies that keep him off the grid. Indeed, the kind of community that has developed around issues of sustainability owes a debt not only to political activist organizations but also the kind of abstract public called into being by such well-established magazines as *Mother Earth News*, perhaps the signal periodical that portrayed a hands-on sustainable life-style and community. If the two broad genres of greenblogging map amateur onto professional, the interchange between the readership creates a virtual community that tries to find ways to deal with the local effect of global industry along with the global effect of local choices.

Yet the bloggers I examined for this study often figure choices as refusals, and it is around refuse that they find the material spaces of everyday life to be most vexing. How, they ask, can people decrease the amount of garbage they produce? How can they keep things out of the waste stream in the first place? How can they actively take things out of the waste stream that have been placed there by others? How can they intervene in larger practices of waste production? Are they just wasting their time blogging about these issues?[7] Green bloggers struggle with these questions, for in some ways garbage demonstrates how completely individual choices as well as refusals are implicated in large-scale systems of commodity production and circulation that seem most resistant to individual action.

Waste is thus the most obvious meeting point of the personal and political. Just as Elizabeth Royte (2005) began her book on garbage by describing exactly what she was throwing away every week, so too do most green bloggers take an inventory of what they discard. They note that most commonly, they discard food scraps and waste (which accounts for 23 percent of household garbage in the United States, according to the Environmental Protection Agency's latest statistics), paper, and packaging materials ranging from cardboard and Styrofoam to plastics. If making waste visible is the aim of so much popular writing on how to live an environmentally sustainable life, it is no surprise that many popular periodicals and books on green lifestyles include discussions of garbage. As Heather Rogers (2005, 3) writes, garbage is "the visible interface between everyday life and the deep, often abstract horrors of ecological crisis."

It is, as the activists and artists who work with it also point out, insistently material and present. It can be seen, touched, and smelled. It can be handled, measured, and tracked. Its disintegration can be calculated, journey to landfills and dumps followed, and effect on socially and politically marginalized populations publicized. Everyone makes garbage, and everyone knows what counts as garbage. Garbage is what something is when it is used up: a scrap of paper on the street, a candy wrapper blowing through a park, a bottle cap in a trash bin, and a garbage bag in a dump.[8] In its material form as the sign of excess consumption and its abstract economic form as the sign of massive systems of global interconnectedness, garbage embodies both how individual consumption produces environmental degradation and the effect of individual action to remedy that degradation. Indeed, as Gay Hawkins (2006) argues, waste is the very sign of how a subject connects to larger structures, how one comes to see oneself implicated in them, and how one comes to experience an "environmental subjectivity." Thus its power is also metaphoric, and the writers of popular environmental books and blogs maintain that paying attention to its material as well as metaphoric qualities can reveal the promise as well as the limitations of individual attempts to comprehend environmental issues in the small compass of one's own life along with the concentrically bigger compasses of one's neighborhood, community, and nation.

Bloggers generate and share strategies for dealing with their waste at the point of both consumption and deciding to get rid of an object. No-ImpactMan, experimenting with zero impact, decided that he and his wife would simply buy nothing new. When they bought food, they only bought those items that were local and had no packaging. They composted their food. They composted their own waste. Beaven has been frank about discussing these choices, but he seems to draw the line at talking about his composting toilet, although among the hundreds of questions and responses are a steady stream of requests from readers who want to know exactly how the composting toilet works, whether it smells, and so forth. These are the sorts of questions to which garbloggers—bloggers who concentrate on waste only—including Leila Darabi at *Everyday Trash: A Closer Look at What We Throw Away* (http://everydaytrash.com), Jonathan Bloom at *Wasted Food*, Dave at *365 Days of Trash*, and *Last Night's Garbage*, a photoblog about garbage in New York City—pay meticulous attention.[9] Two of the biggest green Web sites, Treehugger (http://www.treehugger.com) and Grist (http://www.grist.org), which aggregate environmental news, regularly report on international news, including garbage strikes, dump management, garbology, the culture of dump pickers in developing countries, and the social effects of toxic waste. These omnibus green sites are especially interested in tracing the life cycle of objects—anything from a washing machine or a car to a plastic bag—in order to calculate the effects of producing objects, using them, and throwing them away.[10]

When green bloggers who deal with garbage try to calculate such costs on their own, they tend to keep their calculations within a domestic economy of what their household needs and wants, focusing on the project of deciding how to untangle necessity from desire. In this way, envirobloggers are attuned to both turning their attention to themselves and in their posts, arguing that the small changes they make can add up to a bigger change. They often invoke the rhetoric of labor when they describe their research, and they aren't wrong. It is difficult and time-consuming to discover which places sell locally sourced foods, where they can take broken objects to be repaired, and how they can create workable local communities. But envirobloggers also claim that if they can do it, so can anyone, and by doing so, they are especially good at finding out how to repurpose or recycle various goods, reporting on the local and national places that will take back various used artifacts. (*Everyday Trash*, to take one example, posted information about how to send Crocs back to the show manufacturer—a small triumph, but one in part earned by insistent consumer demand). Discussions of garbage in blogs are not, that is, only about trying to control consumption. They are almost always linked to larger meditations on how we imagine what counts as garbage, how we can intervene in corporate practices, and how individuals can end their complicity in those practices. For many of the bloggers, as I will discuss in the next section, the very act of maintaining a blog is part of their attempt to imagine how they are implicated in various organizations, ranging from corporations to other kinds of green and virtual communities.

III

In this final section, I want to turn from the rhetoric of the blogs I have looked at to examine the form of the blog itself. What, if anything, does the technology of social networking have to do with the technology of environmental self-making that green lifestyle blogs display? The blogs I have discussed are, for instance, indebted to the popularity of books like *Fast Food Nation* by Eric Schlosser (2001), memoirs of experiments in living more locally such as *Animal, Vegetable, Miracle* by Barbara Kingsolver (with Hopp and Kingsolver 2007) and *Farewell, My Subaru* by Doug Fine (2008), and popular analyses of garbage like *Garbageland* by Royte (2005) and *Gone Tomorrow* by Rogers (2005). Although such books contain a great deal of research about the environmental impact of everything from food production to waste management, I note them here because they have important formal and ideological similarities to the blogs I explored.

Like the envirobloggers I look at, these books are organized largely around the difficulties of confronting systemic environmental crisis on an individual level. Whatever large-scale economic analyses each writer makes are framed by and filtered through accounts of personal decisions to live more lightly, consume less, and pay attention to everyday practices. The narratives often contain descriptions of how difficult it can be to find information about the life cycles of the most common household objects—how they are made, distributed, and disposed of—as well as minutely detailed adventures in discovering others who have been struggling to live a more sustainable life. All of the authors talk about how to weigh individual choices against the sense that individual choice might be useless and futile. Writers describe feeling overwhelmed by ignorance about how products they use daily are made and distributed, their inability to fix something that breaks, and their sense of isolation. And perhaps most important, many of them begin with a sense that their efforts are being wasted, even as they are trying to control the wasteful practices that attach to everyday acts of consumption. They are, in other words, like many of the envirobloggers. So how are the blogs different? Are they more interactive than such print books (a question worth asking, given how many blogs are converted into print)? Do they produce public discourse in a virtual green space? What would be the utility of such a green virtual space?[11]

New media and new technologies have played a vital role in the new green movement, uniting the individuals who are practicing what we might call lifestyle activism and more traditional, organized activist groups, and it has helped many individuals to conceptualize an idea of community by letting them link to and work with like-minded individuals. But if the blogosphere aids environmental writers in connecting to communities of like-minded people, it also illustrates the very real difference between a community and audience. Clay Shirky (2008, 85) observes "an audience isn't just a big community; it can be more anonymous, with many fewer ties among users. A community isn't just a small audience either; it has a social density that audiences lack. The bloggers and social network users operating in small groups are a part of a community," even if it is a community we do not immediately recognize as such. Shirky's point, though, should lead us to see the porousness of the distinctions between audiences, at whom information was directed, and traditional environmental communities, which are more immediate, physical, and present. If virtual communities don't seem to have the social density that activists associate with traditional communities, they can be remarkably sensitive to the materiality of objects as well as the geographic locations in which environmental concerns are often first experienced and blogged. Thus, online communities seek to root themselves to the material locations in which bloggers and readers exist, but they also strive to take advantage of knowledge that is frequently not available to local readers and writers, in order to imagine the globe as composed of networked local cultures.

In this sense, envirobloggers work to build community rather than merely audiences; even those environmental online sites structured more like conventional magazines and journals have various reader threads where commenters can critique a given news item or discuss it among themselves. But the sense of community is generally more complex than we have yet understood. Many bloggers understand that they are part of a local and online community, and from these two places, they attempt to assert a sometimes naive sense of global community around environmental issues. As a pedagogical tool, the Internet, especially the blogosphere, is an important resource, despite the charges of some critics that it is flooding popular consciousness with unverified and often vacuous information—an information-age version of garbage. That charge mistakes the very nature of blogging in the first place. Blogs do not pretend to be conventional journalism; they are a far more sociable mode of communication, relying not on the passive absorption of information but instead on the active circulation of ideas in process. Indeed, the self-correction that is now everywhere on the Web has given bloggers the power to both circulate rampant gossip and try to stop it in its tracks. The Internet, as Scott Rosenberg (2009, 9) contends, is also not just about announcing one's opinions to the world at large; it is a tool that can foster collaboration and community.

In *Convergence Culture*, Henry Jenkins (2006, 23) argues that we need to think about two aspects of online communication: the first is access to computers and technology; the second is what he calls participation—which shifts the emphasis from access and technology, to "cultural protocols and practices." Technology itself generates waste, and usually "access" to the technological goods of the rich world is merely access to the broken machines that enable rich-world users to join a global conversation about environmental issues. But it is worth thinking along the lines of Jenkins and other scholars who focus on net knowledge, for these critics are directing our attention to the process whereby communities form and, perhaps more important, the process by which collective intelligence can be brought to bear on crucial issues. Jenkins (ibid., 27) writes that emerging online communities "are defined through voluntary, temporary,

and tactical affiliations, reaffirmed through common intellectual enter-
prises and emotional investments," and that when they work well, they
showcase a "collective intelligence" that can "leverage the combined expertise of their members."

The bloggers I have discussed here, then, are creating a community
that is not, as it might appear when reading them individually, entirely
idiosyncratic. The medium in which they are working cannot ensure a
big audience, nor that each individual blogger will gain more and more
expertise. A blogger who does not engage with comments, post regularly,
include other like-minded blogs on a blogroll, or appear to be participating in a larger conversation or addressing an audience of like-minded
people will join the ranks of most bloggers, and will be read by few, if any,
people outside their immediate circle. The judgment about how popular
a blog becomes is thus not entirely about content but also about the capacity to use the medium to its fullest potential. But those bloggers who
use the medium of the blog to address an audience of like-minded people
they may or may not know will have the opportunity to join an existing
community of bloggers as well as help to create and direct its energies in
different directions.

Jenkins maintains that the collective intelligence that is produced by
net culture is not just the product of the intensive testing and refinement
of ideas in a community of like-minded interlocutors and participants.
It is also about "convergence," or what he describes as the interaction
of user-generated content on the Web with conventional news sources
that produce knowledge in a more top-down way from one expert to
an audience of interested amateurs. The environmental bloggers are in
some sense already part of an older environmental community—one that
had its own publications and news sources, many of which can now be
accessed online. It is not simply the case that environmentalism has gone
mainstream, although that is more or less true. In the case of the environmental bloggers I examine, it is more the case that they push back against
conventional news sources, which must then learn to address them in
ways that they will find compelling.

In terms of the green columns in the *New York Times* and Green Lantern columnists at the online publication *Slate* (slate.com), environmental
reporting has evolved into a hybrid genre that owes much of its energy
to the collaborative thinking that the environmental bloggers represent
at their best. In fact, *Slate* attempted a crowdsourcing experiment, and
called on the expertise of "the hive mind" to develop suggestions for how
Americans might use less energy at home and save money—without really

disrupting their lives. The editors wrote, "You responded with more than 400 ideas, and now we have chosen the top 12, half of them selected based on reader votes, half of them selected by expert judges" (Rogers 2010). This is in line with the most popular lifestyle bloggers I have explored here. Less about the large structural changes that must be made in the production and transmission of energy than about how to make household changes in the consumption of energy, the contest for the hive mind looks decidedly individualistic. But it is also informed by the enormous success of that lively world of environmental bloggers, both in its appeal for suggestions, and its belief that readers will have and want to contribute them, plus will be willing and able to adopt and adapt them.[12]

What, then, does enviroblogging have to teach us about the possibility of public discourse in a world of increasingly atomized, individualistic technologies of self-expression and communication? Certainly, it has much to say about the process by which subjects come to understand their individual passions and commitments as part of a wider network of ideas and action. The blogs I studied are not merely diaries that describe the project of making an environmental self. They are records of how that self comes to see itself as a subject in and through political and social discourses that themselves underwrite popular environmental thinking. They are documents that detail how that subject comes to understand the possibility of its own complicity in the larger systems it is trying to escape, and allow us to see how social actors quite literally link to other individuals and networks that will expand their sense of responsibility to as well as for the ideas that motivated them in the first place.

Lifestyle blogs, like *NoImpactMan*, begin with the desire to purify one's life, and as a result, bloggers frequently discuss how much healthier they feel, how much more money they have, and how much more closely connected to their individual face-to-face communities they feel. The project of self-purification is, in this sense, successful. To use a therapeutic term that has even more traction in this context, envirobloggers often report that they feel more grounded—not merely as a result of their closer attention to their domestic and local environments, but paradoxically also because of their involvement with virtual communities in which they develop affiliations that feel local.

On the other hand, that is a kind of localism that can be provincial; the blogs I look at are also prone to a narrow or narcissistic self-examination even though they focus on larger, planetary green issues. Nonetheless, the "user-generated" content in the blogosphere and the environmental movement is an important way to recognize how local understandings of environmental crisis can be productively used to expand a personal environmental sensibility—one all too frequently prone to self-dramatization or moralization, beyond the immediacy of one's own life, and into a community that would otherwise not be available to an individual.

Zygmunt Bauman (2004, 117) asserts that "liquid modern life," a life characterized by waste,

no longer feels like a culture of learning and accumulation like the cultures recorded in the historians' and ethnographers' reports. It looks instead like a culture of disengagement, discontinuity and forgetting. In such a culture and in the political strategies that it values and promotes, there is not much room for ideals. There is even less room for ideals that prompt a long-term, continuous and sustained effort of small steps leading hopefully toward admittedly distant results. And there is no room at all for an ideal of perfection, deriving all its allurement from the promise of *the end* of choosing, changing, improving.

The story I am telling here is not about how computers will help us save the earth or how envirobloggers will magically be able to transform their aesthetic of self-perfection into the grounds for a vibrant political community. Still, it is worth considering that what used to be understood as fringe movements—blogging and environmentalism—have now become relatively mainstream. Together they are producing communities that have not yet been fully explored because they are not yet fully finished in their ability to account for how the subject can live in a local that seems to transcend prior definitions of that term. Neither have such communities fully found a way to stage a public conversation that can reach beyond the limits of the local that they have established for themselves, although the very form of the blog forces a constant evaluation of how some people get access to the tools to make themselves heard in an emerging public conversation as well as how the local must expand because of their voices. In such communities, in which the technologies of self-improvement and new communicative technologies intersect, an ideal of self-perfection that can seem too closely bound to a blogger's class, regional, and race privilege might also, in some sense, make room for the kind of reflection and participation that can address a wasteful, atomistic, discontinuous way of living in a natural and cultural world in which global and local lives are inextricably intertwined.

Notes

1. Politics by consumption is not, as scholars have shown, merely about making better choices. It can be rooted in a loss of belief in active citizenship—as Maniates (2002) and Princen (2002) variously assert—or it can stem from a desire to

insulate oneself from toxicity by an "inverted quarantine"—as Szasz (2007; 4) argues in a formulation that complements, even if grimly, Foucault's model of self-care—the cordons of which must be purchased and continually maintained by other acts of consumption.

2. Most famously, popular writer Thomas Friedman (2009) argues against the "ten things I can do to save the earth" mentality.

3. Scott Rosenberg (2009, especially chapters 3 and 4) spends a good deal of time on tracing the origins of the "first" blog—a task that is finally nearly impossible given how many different ways blogging evolved from the form of its Internet use in the 1990s. Although now blogging is usually associated with user-generated content, it also, he argues, can be traced to the aggregation of links to other sites on an individual's Web site.

4. *Technorati*, the online journal that tracks the blog and releases an annual "state of the blogosphere report," estimates that as of the writing of this chapter, there are 1,186,653 blogs. Technorati.com is an important site for keeping track of the blogosphere; it tracks the "authority" of blogs by tracking their users, and tracks their rank in their category in the blogosphere. (Here, authority means something more like "influence.") The category of green blogs currently includes 2,108 blogs as of April 2010, though that category includes online journals that might not be understood as blogs, such as *Grist.org*.

5. See http://littlebloginthebigwoods.blogspot.com; http://noimpactman.type-pad.com; http://scienceblogs.com/casaubonsbook; http://www.wastedfood.com; http://www.zerowastehome.blogspot.com; http://www.myzerowaste.com; http://www.cleanbinproject.com; http://www.cheaplikemeblog.com.

6. The rest of the description reads: "EcoGeek devotes its pages to exploring the symbiosis between nature and technology. If you're interested in that, then stop by, and stop by often. . . . EcoGeek is written by Hank Green, Dave Burdick, Phillip Proefrock, Gavin Harper, Matt James, Nino Marchetti, A. Siegel, Billy Shih, Bob Ewing and Jon Schroeder."

7. The reduce/reuse/recycle mantra is important here—and in exactly that order. Generally, recycling programs have been a crucial way for a sympathetic general public to consider environmental issues both structurally and personally, yet as many critics have argued, recycling has become palliative. That is, people believe that merely by recycling the most commonly used household materials—aluminum, plastics 1 and 2, and paper and cardboard—they are helping to "save the planet." But as long-term garbage and green activists know, recycling is the least important way to divert material from the waste stream. It is preferable to reduce use—at the point of production and consumption—and reuse materials before getting them out of the house. For an excellent history of this issue, see Pellow 2007.

8. Jane Bennett (2004) asserts that people can experience ephemeral pleasure by learning to see objects as though they are like contingent art installations. Her primary example is litter. Like Bennett, I believe we need to learn how to see quotidian objects. But if a generous reading of her contention—itself part of a "new materialist" philosophical project—is that Bennett is "repurposing" garbage—that

is, using objects for a purpose for which they were not designed—a more critical claim is that repurposing garbage in order to attain a state of transcendence is in fact a dematerialist project. That state of transcendence is arguably the spiritual version of a deeply socially constructed class and raced ability to "rise above" the contamination of garbage, tying together the twinned meanings of the word refuse around the material sign of garbage.

9. See http://365daysoftrash.com/blogspot.com; http://www.lastnightsgarbage. com. There are a number of specialized blogs that deal only with garbage in the developing world. For example, the Temas Blog (http://www.temasactuales.com/ temasblog/) by Keith R., who describes his trilingual blog as "musings about the evolution of consumer, environmental and health policy in Latin America and the Caribbean," includes a spectacular photo series on garbage dumps and the families who work them.

10. Calculating carbon footprints is inexact, though some manufacturers are now attempting to include those calculations on their products. See Following the Footprints 2011.

11. This chapter does not deal with e-waste, which is a problem that many envirobloggers do not pay much attention to, although it is intimately connected with blogging. Alastair Iles (2004, 76) points out that e-waste contain up to a thousand toxic substances, and itself exemplifies the "pattern of technology and materials flows in the contemporary world economy." Indeed, although the Basel Convention and Basel Ban have tried to regulate the disposal of toxic waste around the globe, especially the trade in toxic waste that disproportionately affects emerging nations with less stringent labor and environmental laws, the problem of e-waste is increasing rather than decreasing, because emerging nations are now hungry for the technology that is now standard in the rich world. For more information on e-waste and the recycling of e-waste, see the Environmental Protection Agency Web site (http://www.epa.gov/). In a global context, the Basel Action Network tracks the use of emerging nations as dumping grounds for the heavy metals and toxic elements of consumer electronics, and carries on its home page a list of articles and reports from media sources around the globe as well as from nongovernmental organizations that are explicitly devoted to combating the trade in toxic waste.

12. That crowdsourcing experiment also produced a "real-time" meeting of two hundred readers and contributors on energy efficiency on March 10, 2010, in Washington, DC.

References

Averett, Nancy. 2010. Museum-Worthy Garbage: The Art of Overconsumption. *Discover Magazine*, April 27. http://discovermagazine.com/photos/27-museum-worthy-garbage-the-art-of-over-consumption (accessed May 25, 2011).

Bauman, Zygmunt. 2004. *Wasted Lives: Modernity and Its Outcasts*. Boston: Polity Press.

Bennett, Jane. 2004. The Force of Things: Steps toward an Ecology of Matter. *Political Theory* 32, no. 3 (June): 347–372.

Fine, Doug. 2008. *Farewell My Subaru: An Epic Adventure in Local Living*. New York: Random House.

Following the Footprints. 2011. *Economist* 399, no. 8736 (June 4): 14–18.

Foucault, Michel. 1988. *The Care of the Self*. Vol. 3, The History of Sexuality. Trans. Robert Hurley. New York: Vintage.

Friedman, Thomas L. 2009. *Hot, Flat, and Crowded: Why We Need a Green Revolution and How It Can Renew America*. 2nd ed. New York: Farrar, Straus and Giroux.

Hawkins, Gay. 2006. *The Ethics of Waste: How We Relate to Rubbish*. Lanham, MD: Rowman and Littlefield.

Iles, Alastair. 2004. Mapping Environmental Justice in Technology Flows: Computer Waste Impacts in Asia. *Global Environmental Politics* 4, no. 4 (November): 76–107.

Jenkins, Henry. 2006. *Convergence Culture: Where Old and New Media Collide*. New York: New York University Press.

Kingsolver, Barbara, with Steven L. Hopp and Camille Kingsolver. 2008. *Animal, Vegetable, Miracle: A Year of Food Life*. New York: HarperCollins.

Maniates, Michael. 2002. Search of Consumptive Resistance: The Voluntary Simplicity Movement. In *Confronting Consumption*, ed. Thomas Princen, Michael Maniates, and Ken Conca, 199–236. Cambridge, MA: MIT Press.

myzerowaste.com/category/blog/.

Newell, Peter. (August 2005). Race, Class and the Global Politics of Environmental Inequality. *Global Environmental Politics* 5 (3):70–94.

Pellow, David Naguib. 2007. *Resisting Global Toxics: Transnational Movements for Environmental Justice*. Cambridge, MA: MIT Press.

Princen, Thomas. 2002. Consumption and Its Externalities: Where Economy Meets Ecology. In *Confronting Consumption*, ed. Thomas Princen, Michael Maniates, and Ken Conca, 23–42. Cambridge, MA: MIT Press.

Rogers, Heather. 2005. *Gone Tomorrow: The Hidden Life of Garbage*. New York: New Press.

Rogers, Jenny. 2010. Twelve Ideas: Choose the Best One. Slate.com, April 12. http://www.slate.com/id/2249915 (accessed May 25, 2011).

Rosenberg, Scott. 2009. *Say Everything: How Blogging Became What It's Becoming and Why It Matters*. New York: Crown Publishers.

Royte, Elizabeth. 2005. *Garbageland: The Secret Trail of Trash*. New York: Little, Brown and Company.

Schlosser, Eric. 2001. *Fast Food Nation*. New York: Houghton Mifflin.

Shirky, Clay. 2008. *Here Comes Everybody: The Power of Organizing without Organizations*. New York: Penguin Press.

Szasz, Andrew. 2007. *Shopping Our Way to Safety: How We Changed from Protecting the Environment to Protecting Ourselves*. Minneapolis: University of Minnesota Press.

II

The Places of Garbage

4

Missing New Orleans: Tracking Knowledge and Ignorance through an Urban Hazardscape

Scott Frickel

Any society's relationship to the land—how people treat the dirt beneath their feet—is fundamental, literally.

—David. R. Montgomery

Dirt and waste are close cousins.[1] Or are they? Although we may consider them roughly interchangeable as linguistic tropes, the ontological status of dirt and waste are worlds apart. If waste is inert matter, dirt—or more properly soil—teems with life; it is "a dynamic system that responds to changes in the environment" (Montgomery 2007, 13). Indeed, "waste" that contains life, as in my backyard compost bin, is not waste at all but rather a part of this dynamic system, on its way to becoming dirt. If waste is that which is little valued, moreover, dirt is perhaps humanity's most precious natural resource—what William Bryant Logan (1995, 14) describes as earth's "ecstatic skin," the barrier between bedrock and atmosphere that makes life on this planet possible, and "from which all life on the land is born." Further still, if waste marks the end of the industrial production line, dirt—or the stones, clays, minerals, and fossil fuels that in time will become dirt—marks industrial production's extractive beginnings. Finally, if waste tends to concentrate, for example in dustheaps, leach ponds, or landfills, dirt is without exception everywhere, quite literally the ground beneath our feet.

There are two key moments when the ontological distinction I am drawing between dirt and waste breaks down. One moment occurs with composting, when waste, is transformed into dirt. The other happens with the contamination of soil by environmental toxins, when dirt is transformed into waste. I view such instances of ontological ambiguity as opportunities for critical inquiry into the historical relationships entwining ecological and social systems. There are many ways to take advantage of such opportunities.

This chapter addresses the problem of urban soil contamination. More specifically, it examines how regulatory agencies use science to make waste from dirt. The perspective I take on this topic is decidedly more epistemological than ontological. I am less concerned with waste per se than with knowledge about waste. So rather than ask what contaminated soil is and how it got that way, I instead ask how regulatory agencies produce knowledge about soil contamination, and how they use and disseminate that knowledge to address public concerns about environmental risk. And because the production of knowledge always involves the simultaneous production of ignorance, I also ask about the forms that ignorance takes in regulatory science, how the lack of knowledge informs regulatory decisions regarding the presence or absence of contaminated soil, and what, if anything, should be done about it.[2]

My points of departure are two recently completed studies of the "hazardscape" in New Orleans, Louisiana, prior to and following the landfall of Hurricane Katrina in 2005.[3] Both examine the problem of urban soil contamination, but from different angles of approach. The first study comes at the problem historically and investigates relict industrial waste.[4] A main goal of this work is to identify the locations of former industrial facilities across the city. Over time, some of those facilities have converted to other land uses, and such conversions may effectively hide the historical accumulation of toxins that remain on those sites. The study seeks to recover that lost knowledge. The second study is more contemporary, and investigates the uneven distribution of knowledge and consequent production of knowledge gaps resulting from the hazard assessment by the US Environmental Protection Agency (EPA) of New Orleans soil and flood sediment in Hurricane Katrina's aftermath.[5] The study disassembles the EPA's assessment data to map the spatial distribution of knowledge and identifies several knowledge gaps—neighborhoods in which no knowledge about soil contamination was ever produced. I use the findings from these paired studies to consider how lost knowledge of historical hazards and contemporary knowledge gaps generated by regulatory agencies are related spatially, and how these two ways of missing New Orleans can inform a deeper appreciation of—and concern for—the historical nonproduction of environmental knowledge along with the subsequent institutionalization of environmental ignorance.

Environmental Knowledge/Ignorance

An emerging thesis in the social studies of science is that the scientific production of knowledge and ignorance are intertwined and often countervailing processes. This knowledge/ignorance thesis challenges classic studies of scientific and medical knowledge by historians, philosophers, sociologists, and others who limit their analysis to the construction of facts and practices of knowledge making.[6] It also challenges earlier sociological work framing ignorance as a more or less stable background condition that poses particular sorts of problems for decision makers confronted with uncertainty, doubt, or inaccurate information.[7] Rising to both challenges is a stream of recent work that presents ignorance as culturally and historically specific outcomes deeply entwined with knowledge making, and every bit as complex and context dependent.[8]

Different scholars come at the problem of knowledge/ignorance in different ways. Some depict ignorance as the result of purposive or strategic action, with scholars investigating how certain kinds of knowledge are purposefully made inaccessible to others, for instance, through secrecy and censorship, deceit and suppression, and research agenda setting.[9] Such studies present ignorance as intentional outcomes, driven more or less directly by political, economic, or other social interests. Other scholars depict ignorance as the largely unintentional or derivative results of longer-term historical and cultural processes. Londa Schiebinger's analysis (2009) of the disappearance of knowledge about West Indian abortifacients or Linda Nash's efforts (2007) to document the winnowing of environmental knowledge from medical understandings of disease are good examples of the ways in which once-prominent forms of knowledge become lost or eroded. I follow many of the insights gained from this recent wave of scholarship to investigate the social production of knowledge/ignorance about soil contamination in New Orleans.

Study One: Lost Knowledge

Taking a cue from research by environmental historians and historical geographers on the accumulation and disposal of relict industrial waste, the first study investigates the spatial distribution of industrial lots in New Orleans along with the conversion of those parcels to other commercial and noncommercial uses over a fifty-year period (1955–2006).[10] My goal in this research, done with my colleague James R. Elliott, is to understand what sites of hazardous industry become over time. When manufacturing facilities close or change owners, do hazardous land uses tend to continue on those sites? Or do former sites of hazardous industry instead tend to convert to other, more environmentally benign land uses such as parks and playgrounds, retail outlets, or residential housing? By investigating how a city's industrial footprint changes spatially and over time, we can begin to understand the dynamics by which hazardous wastes accumulate in cities, and under what conditions those wastes become "hidden" unintentionally through larger-scale urban processes such as deindustrialization, out-migration, and gentrification. An unintended consequence of such land use conversions is that community knowledge about past hazardous processes may be forgotten with time, and thus rendered inaccessible to city residents, urban planners, and environmental regulatory agencies for whom accumulations of (now-invisible) hazardous waste goes unrecognized.

For this study, we identified six industrial sectors that have had a significant economic presence in New Orleans during the study period—chemicals, petroleum, plastics, primary metals, stone/clay/glass (mostly concrete), and transportation equipment (mostly involving shipbuilding and repair)—and for which prior research confirms a high probability of on-site contamination (ranging from 83 percent for primary metals to 95 percent for plastics manufacturers).[11] Next, we collected historical information on specific facilities by consulting the *Louisiana Directory of Manufacturers* for 1955, 1965, 1975, and 1985. These directories are organized by city and industrial sector. They provided us with New Orleans street addresses as well as data about a firm's name, the number of its employees, and Standard Industrial Classification product codes. This approach netted 215 unique addresses. We then selected 94 of those addresses at random and conducted site surveys to identify the type of land use on each site today (such as, for example, hazardous manufacturing, vacant or underused lots, commercial operations, residential units, or public spaces), consulting company Web sites and local business directories as necessary to confirm or correct our visual field observations. Finally, we merged our historical manufacturing and contemporary site survey data with US Census data (1970–2000).

By examining site-specific changes between earlier periods of post–World War II urban development (t_1, defined as 1955–1985), and the more recent period of redevelopment (t_2, which runs from 1986 to 2006), we can begin to identify organizational, neighborhood, and urban influences on land use conversions that may potentially hide—but not erase—historically accumulated industrial waste. We can also compare our list of 215 historical industrial sites to lists of hazardous waste sites and industrial brownfields developed by regulatory agencies. A large overlap would indicate that existing regulatory mechanisms do a reasonable job of identifying potential relict industrial waste sites in New Orleans; a small overlap would suggest that relict industrial waste in the city is a more substantial problem, and would speak to the environmental policy and public health implications of recovering lost knowledge of the city's industrial past.

New Orleans' past as an old port city is a storied one, and includes a long history of manufacturing activities. Despite this history, New Orleans has never been a center for heavy industrial production. So it is not surprising that our data collection efforts identified only 215 historical industrial sites, as shown in table 4.1. This is a relatively small number for a large US city and speaks to the limited intensity of New Orleans' industrialization historically. By comparison, in Portland, Oregon, another historical port city with roughly the same population size as pre-Katrina New Orleans, we identified 714 unique industrial sites using the same methods.[12]

If the scope of industrialization was relatively small in New Orleans, so was the scale of industrial activities. Most of the sites we identified were occupied by relatively small operations averaging between ten and seventy employees that had clustered mainly, although not exclusively, along the city's water, road, and rail corridors. Small or not, these manufactories likely packed a hefty environmental punch. Much of the city's industry after midcentury was tied to the region's oil and natural gas resources, with the timing of that development coming on the heels of an oil boom that crested in the late 1970s.[13] Over half of the facilities in our sample either processed oil and natural gas into petroleum, or processed petroleum into chemicals or plastics; many of the rest supplied the oil fields with marine transportation equipment and pipeline hardware, or furnished the accompanying construction boom with concrete and other building materials. Based on information contained in the Historical Hazardous Substance Data Base (Colten 1992), it is probable that dozens of persistent environmental contaminants may have been used for these and associated industrial activities during the past fifty-plus years. And since manufacturing is an inherently messy business, many of those contaminants were likely to have been buried, dumped, injected, spilled, or rinsed, or otherwise came to contaminate the soils at an unknown number of these sites.

This is even more likely given that most of these facilities were older, thereby operating well before state and federal agencies provided substantial regulatory oversight of hazardous waste production and disposal, and during a time when New Orleans' policies regulating industrial zoning and waste disposal were underdeveloped as well as unevenly enforced.[14] We located two-thirds of these sites (n = 142) in the earlier 1955

Table 4.1
Hazardous industrial sites, by directory year of initial observation*

	Directory year of initial site observation				Total	
	1955	1965	1975	1985	#	(%)
All sites (in descending order of industrial frequency within city)						
Chemicals	21	36	11	11	79	(36.8%)
Stone/clay/glass	10	14	9	6	39	(18.1%)
Transport equipment	12	10	11	5	38	(17.7%)
Primary metals	19	5	2	3	29	(13.5%)
Plastics	6	6	3	10	25	(11.6%)
Petroleum	2	1	2	0	5	(2.3%)
Total #	70	72	38	35	215	(100%)
(%)	(32.5%)	(33.5%)	(17.7%)	(16.3%)	(100%)	
Sampled sites (in descending order of industrial frequency within city)						
Chemicals	11	15	6	3	35	(37.2%)
Stone/clay/glass	5	5	6	1	17	(18.1%)
Transport equipment	6	3	5	1	15	(16%)
Primary metals	7	1	1	3	12	(12.8%)
Plastics	1	5	2	5	13	(13.8%)
Petroleum	1	1	0	0	2	(2.1%)
Total #	31	29	20	13	94	(100%)
(%)	(30.8%)	(28.8%)	(20.2%)	(20.2%)	(100%)	

Source: Adapted from table 1 in Elliott and Frickel 2011.

* Our primary unit of analysis is sites, or addresses. A site is counted only once for the first directory year in which it is observed operating in one of the six specified industries. The industrial sector is recorded from this point of first observation. If an establishment moves or expands to a new location between directory years, the new site is counted when it is first observed in operation at the new address.

and 1965 manufacturing directories, and only half that many in the more recent 1975 and 1985 directories. This suggests that for a large majority of sites in our study, conversion to other land uses could have occurred any time during a forty- to fifty-year period prior to conducting our site surveys in 2006. And those conversions could have taken place any time during a fifteen- to twenty-five-year period prior to the passage of the Comprehensive Environmental Response, Compensation, and Liability Act (aka "Superfund") in 1980 requiring the EPA to identify hazardous waste sites and making the owners of contaminated properties legally liable for cleaning them up.[15]

When we visited 94 of these former industrial sites in summer 2006, we found more than a quarter (28.6 percent) to be still occupied by manufacturing establishments, and another 12 percent to be vacant or underused lots. Totaling just over 40 percent of our sample, currently operating industrial facilities and derelict brownfields are the two endpoints that have attracted the most attention among academics studying urban industrial hazards and environmental justice.[16] Yet a large majority of sites in our sample (59.4 percent) had converted to various nonindustrial uses, be they commercial sites such as restaurants or grocery stores, public and quasipublic uses such as parks, public housing, or churches, or private residences. These are not the endpoints that research on brownfields and environmental justice typically capture, in large part because places like playgrounds or restaurants tend not to be listed on federal and state hazard inventories.

To be sure, we compared our list of 215 historical industrial sites to the state of Louisiana's Voluntary Remediation Program Sites List for Public Record (http://www.deq.state.la.us/portal/tabid/269/Default.aspx). That list was established in 2002 when the state initiated its brownfields identification and remediation program, but contains information that is historically cumulative. This public list of hazardous sites includes only sites where the property owners (or potential owners) sought to obtain a release of liability from the state after cleaning up historical contamination on-site. Even so, it is a remarkably short list, containing only 18 sites located in New Orleans. None of the sites on the state's list overlap with the 215 historical industrial sites we found.

Those sites, and whatever contaminants that may remain behind, represent lost knowledge of various sorts: community knowledge about daily life in and around the facilities, and about the people who worked inside or played nearby; managerial knowledge about the social organization of industrial production in those places; technical knowledge about the materials used there, how they were transformed, and where they went; geographic knowledge about the spatial distribution of those now-relict wastes; and not least, chemical knowledge about how substances change over time as well as in their interaction with air, soil, water, and living organisms.

While it is rarely missed, lost knowledge matters. Because people (and bureaucracies) tend to make decisions on the basis of what they know, rather than what they do not know, lost knowledge limits the possibilities of social action. It is in this sense that lost knowledge can forestall even well-meaning efforts to understand, for example, the relationship between environmental hazards and public health in New Orleans, as our second study illustrates.

Study Two: Uneven Knowledge and Knowledge Gaps

On August 29, 2005, storm surges produced by Hurricane Katrina caused massive failure of the federal levee hurricane-protection system surrounding the city of New Orleans. Overtopping and breeches in the levee walls filled the city with an estimated 131 billion gallons of water.[17] At maximum stage, the flood covered 80 percent of the city's land area and inundated the houses of more than 60 percent of the city's residential population, soaking some neighborhoods for nearly six weeks.[18] The deluge contained a complex mixture of chemicals, minerals, heavy metals, and biological pathogens. The nightmarish risk scenario demanded—and received—an unprecedented response from state and federal agencies tasked with assessing the storm's ecological and human impacts.

Led by the EPA and the Louisiana Department of Environmental Quality (LDEQ), the hazard assessment that followed in the flood's wake was an organizationally massive undertaking. The project involved more than 1,600 EPA staff and contract employees developing as well as carrying out work plans in collaboration with a dozen state and local agencies and departments. The hazard assessment spanned an entire year and produced some 400,000 analytic tests, examining about 1,800 sediment and soil samples for the presence of 200 toxic substances. While regulatory scientists collected data from across hurricane-impacted Louisiana, more than half of those samples—951—came from the flood print inside New Orleans.[19]

On average each sample contained 23 different compounds, demonstrating that New Orleans' soils contain diverse combinations of chemicals and other toxicants.[20] Yet despite the relative diversity of toxicants found in New Orleans soils, contaminant concentrations in most neighborhoods registered below state regulatory standards for long-term human health risks. These findings meant that regulators were not required by law to take further action in those affected neighborhoods.

The main exception to regulatory inaction involved tests for a few contaminants—primarily lead—which were discovered all around the city and in many neighborhoods at levels far exceeding state health risk standards (Solomon and Rotkin-Ellman 2006). But because earlier studies of lead levels in New Orleans soils, conducted by New Orleans–based soil lead expert Howard Mielke (1999), showed that lead contamination predated the hurricane and flooding, EPA officials determined that the agency was not responsible for health risks posed by the city's soil lead problem. In its final report, released just prior to the first anniversary of the hurricane's landfall, the EPA stated that "the sediments left behind by the flooding from the hurricanes are not expected to cause adverse health impacts to individuals returning to New Orleans" (EPA 2006, n.p.).

In the end, a lot of knowledge about the nature and geography of soil contamination translated into almost no remediation of contaminated soils by federal or state agencies. Why? It may be because inferring environmental risk—or in this case, its relative absence—from what is known about contaminants along with their spatial and chemical concentrations is standard regulatory procedure in the United States. As our research is beginning to show, though, whether in response to catastrophic dislocations of social and ecological systems (as in Katrina), or the course of daily regulatory business such as monitoring hazardous landfills or investigating accidental releases, standard regulatory procedure often fails to account for what remains unknown.

We have been a bit less orthodox in our approach, employing the EPA test data to investigate how, where, and when the EPA did and did not produce knowledge about soil contamination. Where our first study sought to recover lost knowledge from New Orleans ground once occupied by manufacturing facilities, this study examines knowledge distributions and gaps produced by the uneven regulatory response following the 2005 flood.[21]

The study measures the spatial, temporal, and epistemological organization of the EPA's knowledge investments during the hazard assessment process, although here I focus solely on the spatial analysis. The term "knowledge investments" refers to the various resources that the EPA expended in generating knowledge for the hazard assessment project. These investments included collecting soil and sediment samples, returning to sampling locations for follow-up sampling, and performing tests on the sampled material. Because samples played such a central role in this process, we count them as the principle form of knowledge investment.

EPA and LDEQ regulators based the hazard assessment on the test results for each soil and sediment sample they collected. In this process, the samples functioned as aggregations of knowledge that "spoke for" soil quality in a given block or neighborhood. The density of knowledge investments, in this context, offers a measure of the proportionate representation or "voice" that samples provide to nearby blocks (and implicitly to people living on or returning to those blocks). It is also a way to assess the aggregate weight of knowledge that samples carry in relation to other samples. Areas in the flood zone that received lots of sampling had more voice and carry more weight in the process than those that received little or no sampling.

We began to investigate the spatial distribution of the EPA's knowledge investments by identifying New Orleans city blocks falling within or overlapping with the flood zone's perimeter. This is a conservative strategy conforming to the EPA's stated mission to assess contaminants only in flooded areas of the city. The flood zone contained 7,231 blocks, which before Katrina had been home to 359,470 city residents. Next, using US Census data, we identified flooded blocks by racial composition and median household income—common variables in studies of environmental inequality in US metropolitan areas.[22] Finally, we plotted the location of the 951 samples collected in New Orleans during the hazard assessment sampling points using publicly available data from the EPA Hurricanes Katrina and Rita Response Project (http://www.epa.gov/katrina/index.html). Then for each racial and income category, we tabulated the total number of sampling points to derive a spatial distribution of knowledge investments across the flood zone in relation to the flood-print population's racial and socioeconomic geography.

Table 4.2 depicts the results of our density analysis for race and median household income. For ease of interpretation, ratios of samples-to-blocks and samples-to-block groups are reported beneath the means. Also, readers should be aware that the city's high levels of residential segregation and highly skewed income distribution demand a more nuanced interpretation than I can provide here.[23] Despite these caveats, the general pattern is clear: samples—and thus knowledge investments—cluster heavily in blocks that are predominantly African American and in neighborhoods that are predominantly poor. The samples, conversely, are most dispersed in areas that are predominantly nonblack (that is, white) and least poor.

Table 4.2
Spatial distribution (density) of knowledge investments by race and median household income

	Race (% black)					Median household income (MHI)			
	Number of blocks (%)	Number of sample points (%)	Meanb (sample-to-block ratio)	Standard deviation	MHI block groups	Number of block groups	Number of sample points (%)	Meanb (sample-to-block group ratio)	Standard deviation
Zero population blocksa	1,955 (27)	119 (12.5)	0.058 (1:17.2)	0.380					
% black blocks									
< 20%	1,062 (14.7)	74 (7.8)	0.071 (1:14.1)	0.414	$47,842– 200,001	178	91 (9.6)	0.511 (1:2)	2.017
20–39%	262 (3.6)	19 (2.0)	0.069 (1:14.5)	0.268	$35,589– 47,841	178	111 (11.7)	0.624 (1:1.6)	2.246
40–59%	354 (4.9)	45 (4.7)	0.138 (1:7.2)	0.792	$26,980– 35,588	178	113 (11.9)	0.635 (1:1.6)	2.338
60–79%	571 (7.9)	50 (5.3)	0.084 (1:11.9)	0.350	$18,985– 26,979	179	212 (22.3)	1.184 (1:0.8)	3.004
80–100%	3,027 (41.9)	644 (67.7)	0.213 (1:4.7)	0.702	$1–18,984	179	424 (44.6)	2.374 (1:0.4)	7.176

Sources: Adapted from tables 2 and 3 in Frickel, Campanella, and Vincent 2009; 2000 US Census, block-level population counts for Orleans Parish, LA; EPA Hurricanes Katrina and Rita Response Project.

Notes: a. Includes blocks, parks, streets and medians, traffic circles, industrial areas, and minor anomalous polygons with no residential population. There are no zero-population block groups in New Orleans. b. Mean for density calculated as number of sample points/number of blocks or block groups. Because differences between means are small, to ease interpretation we also report inverted sample-to-block/block-group ratios that describe the number of blocks or block groups "represented" by each sample point.

On average, the EPA collected one sample for roughly every five blocks in neighborhoods whose resident population in 2000 was 80 to 100 percent black, but only one sample for about every fifteen blocks with less than 40 percent black residents. Put another way, the soil and sediment samples that the EPA collected in mainly black neighborhoods carried three times more knowledge than those gathered in mainly white neighborhoods. We find a similar pattern for median household income measured at the block-group level. Roughly speaking, neighborhoods at the bottom of the flood-print population's income distribution received more than four times the knowledge investment than neighborhoods at the top.

Taken at face value, these findings appear to directly contradict the concerns of environmental justice scholars and activists who charge that regulatory agencies disproportionately underinvest resources when investigating environmental hazards in minority and poor neighborhoods.[24] In this case, however, the EPA's overinvestment in one New Orleans neighborhood was in part due to the agency's prior knowledge of environmental health problems related to the old Agriculture Street Landfill. This municipal landfill was closed in the 1950s, reopened briefly in 1965 as a depository for storm debris from Hurricane Betsy, and then covered with topsoil and ash from city incinerators before commercial and low-income residential housing development began on the site in the mid-1970s. When health problems started to appear among the new residents a decade later, the community organized to demand environmental justice. In response, the EPA embarked on a series of environmental investigations, finally placing the site on the National Priority List (the Superfund) in 1994; a remediation program to remove contaminated topsoil was completed in 2002.[25] Because of this history, after the flood the Agriculture Street Landfill site attracted considerable attention from regulators, who collected well over a hundred samples from this approximately eight-block site. Thus, unlike others parts of the city, preexisting hazards buried beneath a single residential subdivision occupied almost entirely by low-income African Americans helps to explain why knowledge investments tilt heavily toward low-income minority areas. I will return to this point later.

In addition to assessing whether some neighborhoods received more knowledge investments or less, we can also say something about neighborhoods that received no knowledge investments whatsoever. In mapping the location of the 951 sample points, we identified three large contiguous areas within the New Orleans flood print where no soil or sediment samples were collected or tests conducted. We view these areas as literal knowledge gaps.[26] These knowledge gaps are spatial in nature, comprised of more than 450 city blocks. They cover densely populated residential areas that are home to nearly a fifth of the flood zone population. Based on US Census data for 2000, these gaps—spanning census tracts with pre-Katrina populations of African American residents above and below the citywide average as well as with the proportion of residents living below the poverty level ranging from 10 to 56 percent—illustrate some of the city's racial and socioeconomic diversity.[27]

Unlike other parts of the flood zone, these knowledge gaps are adjacent to or intersected by industrial corridors and commercial districts. Yet despite their geographic proximity to likely industrial hazards, knowledge gaps in New Orleans are distinguished by the complete absence of sampling and hence the formal nonproduction of spatially relevant knowledge. Where the lost knowledge described earlier resulted from a combination of general historical processes such as deindustrialization and suburbanization, these spatial knowledge gaps are social products of a different sort—the (presumably) unintended consequences of bureaucratized regulatory science.

Between Lost Knowledge and Knowledge Gaps

Created at different times and resulting from different processes, lost knowledge and knowledge gaps are nevertheless related, at least in spatial terms. The map in figure 4.1 shows all 215 former industrial site locations from our first study along with all 951 EPA sediment and soil sample points examined in our second study. While there are exceptions, the general lack of spatial correspondence between potentially contaminated land and knowledge about contaminated soils is clear: historical industrial sites (identified in figure 4.1 as Xs)—those urban parcels most likely to contain legacy contaminants—are located precisely where sediment and soil samples (identified as black circles in figure 4.1) undertaken by the EPA in the name of public health did not occur. By my rough but conservative estimate, the EPA sampling efforts occurred within several blocks of fewer than twenty historical industrial sites, representing less than 10 percent of the total. The vast majority of the EPA sample sites, in other words, come nowhere close to the historical locations that mark the city's manufacturing past and toxic present. While most of these historical industrial sites are located outside the Katrina flood zone and thus lay beyond the (political) parameters of the EPA assessment effort, twenty-seven of the industrial sites identified in study one do fall within the rough visual boundaries of the three knowledge gaps identified in study two. Moreover, because our sample of historical industrial sites is just that—a sample of only six industrial sectors (among dozens) captured during four years (among decades) —there is little doubt that our data underestimates the disjuncture between contemporary regulatory efforts to assess environmental risk and likely historical sources of that risk.

This striking disconnect is explained in part by the EPA's institutional mandate following the hurricanes to assess environmental risk in flooded residential areas. The mandate relieved the agency of its responsibility for investigating neighborhoods that lay beyond the flood zone as well as nonresidential areas such as industrial corridors that intersected or lay within the flood zone. Whether that plan was in the best interests

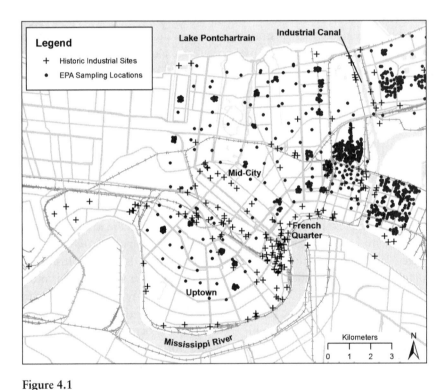

Figure 4.1

Locations of historical industrial sites and EPA sample points, New Orleans
Sources: The EPA sampling points are derived from an analysis of test results data for Orleans Parish, http://www.epa.gov/katrina/testresults/index.html. Industrial site data are taken from the *Directory of Louisiana Manufacturers* for 1955, 1965, 1975, and 1985.

of public health seems debatable, resting as it does on a logic that only loosely conforms to recent history.

Industry in New Orleans has historically concentrated along the Mississippi River, and since the 1930s also along the Industrial Canal that links the river to Lake Pontchartrain.[28] Levees bordering both waterways were at elevations sufficient to protect those facilities from Katrina's flood. Even so, like the rest of the city these industrial zones suffered extensive wind damage during the hurricane and damage from unchecked fires in the months afterward. If one were searching for environmental hazards in New Orleans, these former and contemporary industrial areas would be among the obvious places to look. Incentives for seeking that knowledge are even greater given that since Katrina, the city's relatively high ground along the river has become prime real estate for commercial and residential redevelopment.[29]

Industrial sites at lower elevations located along interstate and rail lines in the city's interior were also, by contrast, systematically avoided in the hazard assessment process, even though these areas did experience catastrophic flooding. And while these interior neighborhoods are industrial, they are not exclusively so. Prior to Katrina, nearly sixty-six thousand New Orleanians were living near existing or former industrial facilities in areas that are best characterized by mixed residential, commercial, and industrial use. These neighborhoods gained new significance in spring 2007 when the mayor's Office of Recovery Management announced that it would target some of these same areas for redevelopment investments because the economic and cultural diversity in such mixed-use neighborhoods was seen as offering a way to anchor recovery efforts.[30] So not only were these interior industrial areas populated before the storm, in 2010 they were well on their way to becoming centers of concentrated repopulation.

Conclusion

What lessons can we draw from these studies of the ways in which knowledge about waste is made, lost, and left undone? The first is a material one: Dirt may be everywhere, but waste is not; waste concentrates. Despite global flows of hazardous chemicals across the planet, toxic substances still concentrate in some environments and communities more than others.[31] Louisiana, for example, is not only a top producer but also a net importer of other states' hazardous waste.[32] Structural inequalities such as this matter, especially when posed against currently fashionable arguments by scholars like Ulrich Beck (1992, 36), whose now-famous admonition that "poverty is hierarchic, smog is democratic" implies that wealth no longer can purchase freedom from technologically introduced environmental risks. Such narratives, while noteworthy, tend to universalize industrial technoscientific power in conjunction with its consequences in ways that mask deepening environmental and social inequalities. Doing so unnecessarily mystifies the political and economic forces that drive the extraction, production, and global spread of synthetic and naturally occurring contaminants. While the wastes generated as by-products of industrial technoscience now seem to be virtually everywhere, they are not evenly or randomly so. This lesson in the materiality of waste suggests two additional lessons that are more epistemological, and involve how we know what waste is and where our ignorance of waste remains.

If hazardous waste is distributed unevenly, our collective knowledge about those hazardous wastes is produced unevenly as well. Industrial chemists and engineers know a lot about making chemicals, but industrial toxicologists and others working in related health sciences know comparatively little about those chemicals' ecological and health effects. The magnitude of this knowledge-based disparity becomes apparent in regulatory contexts. As noted in a US Government Accounting Office (2005) report, the 195 chemicals that testing sought to identify in New Orleans soil and sediment represents less than a quarter of 1 percent of the more than 82,000 substances listed on the EPA's chemical inventory. The report indicates that this is about the same proportion of chemicals for which the EPA is reported to possess "complete health data." Indeed, the deeper one digs into the EPA's regulatory databases, the more thin and sparse the knowledge becomes.[33] These knowledge disparities are institutionalized in hazard and risk assessment, and help to define the conditions under which regulatory agencies treat dirt as waste. In New Orleans, our research shows that similar inequalities in the deployment of (uneven) knowledge also generated spatially specific knowledge gaps and failed to recover lost knowledge.

While they may not be specifically planned, lost knowledge or knowledge gaps do not simply happen. These two different ways of producing ignorance are the results of historical processes that take specific social and spatial form in New Orleans, although they are by no means specific to that city. More important, as our research demonstrates, ignorance itself accumulates spatially, temporally, and institutionally. The absences that it creates through accumulation are complex, often invisible, and typically difficult to measure. In New Orleans, regulatory action—and inaction—propelled by these and other forms of institutionalized ignorance has come at the high price of missing an unprecedented opportunity.

All urban soils contain industrial and other contaminants. But unlike most city dwellers elsewhere, state and city officials overseeing New Orleans' recovery from Hurricane Katrina were uniquely positioned to address soil contamination problems on a scale and in a manner that few officials from other cities have ever had. The hurricanes and flooding destroyed or severely damaged hundreds of thousands of homes and businesses. These along with tens of thousands of less severely damaged buildings and most of the city's public spaces—parks, golf courses, schools, and street medians—are being renovated, rebuilt, redeveloped, or redesigned as green space. In short, the slow-moving recovery has presented the city, the state of Louisiana, and the nation with a historic opportunity to enhance public health and environmental well-being by making an urban ecosystem less toxic.

In the seven years since the storm, experts and activists have put forward several soil remediation plans. The best of these plans sketched rough blueprints for large-scale remediation—either through various bioremediation strategies, topsoil removal, or topsoil replacement.[34] For different reasons (the details of which remain unclear), and despite the concerted and valiant efforts of several environmental and community activists to develop small-scale pilot remediation projects, this historic opportunity has largely slipped by. If, as the regulatory agencies claim, city soils were no more contaminated after Katrina than they were before, they are contaminated nonetheless and remain so today. This, in part, is what it means to miss New Orleans.

Notes

1. Portions of this chapter originally appeared in Frickel 2008. The resources for the original research reported in this chapter were provided by grants from the Social Sciences Research Council and the National Science Foundation. Richard Campanella, James R. Elliott, and M. Bess Vincent collaborated on those studies. Michelle Totman created the map. I thank them all for their contributions, and I thank this volume's editors for constructive comments.

2. See, for example, Gross 2010.

3. See Cutter 2001.

4. See Frickel and Elliott 2008; Elliott and Frickel 2011.

5. See Frickel, Campanella, and Vincent 2009; Frickel and Edwards 2010; Frickel and Vincent 2010.

6. See, for example, Collins 1992; Latour and Woolgar 1986; Shapin and Schaffer 1985.

7. See Smithson 1990.

8. See Gross 2010; Tuana and Sullivan 2007; Proctor and Schiebinger 2008.

9. See Galison 2009; Hess 2007; Markowitz and Rosner 2002; Michaels and Monforton 2005.

10. For an example of research on relict industrial waste, see Colten 1990. On the first study, see Frickel and Elliott 2008; Elliott and Frickel 2011.

11. See Noonan and Vidich 1992.

12. See Elliott and Frickel 2011.

13. See Freudenburg and Gramling 1994.

14. See Tarr 1996; Colten 2005.

15. See Hird 1994; Leigh and Coffin 2000.

16. For overviews, see, respectively, Lee and Coffin 2000; Brulle and Pellow 2006.

17. See Smith and Rowland 2007.

18. See Campanella 2007a.

19. For the EPA test data, see http://www.epa.gov/katrina/testresults/index.html. Metropolitan New Orleans includes contiguously populated portions of Jefferson, Orleans, Plaquemines, and St. Bernard parishes. The analysis summarized here is based on data from Orleans Parish (city of New Orleans) only.

20. See Frickel and Edwards 2010.

21. See Frickel, Campanella, and Vincent 2009.

22. See, for example, Saha and Mohai 2005; Downey 2007.

23. For more details, see Frickel and Edwards 2010.

24. Such as, for example, O'Neil 2007; Cline and Davis 2007.

25. The case is covered in Timmons and Toffolon-Weiss 2001, chapter 6; Colten 2005, chapter 4.

26. See Frickel and Vincent 2010.

27. There were 65,962 residents (18.5 percent) in these knowledge gap neighborhoods and of those, 47,310 residents (71.7 percent) were African American. These racial and socioeconomic characteristics are roughly proportionate to the composition of the larger flood zone population. See Campanella 2007a.

28. See Colten 2005.

29. See Campanella 2007b.

30. See the City of New Orleans 2007.

31. See Pellow 2007.

32. See Timmons and Toffolon-Weiss 2001, 16–22.

33. See Frickel and Edwards 2010.

34. See, for example, Mielke et al. 2006.

References

Beck, Ulrich. 1992. *Risk Society: Toward a New Modernity*. London: Sage.

Brulle, Robert J., and David Pellow. 2006. Environmental Justice: Human Health and Environmental Inequalities. *Annual Review of Public Health* 27:103–124.

Campanella, Richard. 2007a. An Ethnic Geography of New Orleans. *Journal of American History* 94 (3): 704–715.

Campanella, Richard. 2007b. Geography, Philosophy, and the Build/No-Build Line. *Technology in Society* 29:169–172.

City of New Orleans, Mayors Office of Communications. 2007. City Announces First 17 Target Recovery Zones: Areas Will Attract Investment, Residents to Key Resources. Press release, March 29.

Cline, Kurt D., and Charles Davis. 2007. Assessing the Influence of Regional Environmental Protection Agency Offices on State Hazardous Waste Enforcement Decisions. *Social Science Journal* 44:349–358.

Collins, Harry M. 1992. *Changing Order: Replication and Induction in Scientific Practice*. Chicago: University of Chicago Press.

Colten, Craig E. 1990. Historical Hazards: The Geography of Relict Industrial Wastes. *Professional Geographer* 42:143–156.

Colten, Craig E. 1992. *Historical Hazardous Substance Data Base*. Springfield: Illinois State Museum.

Colten, Craig E. 2005. *An Unnatural Metropolis: Wresting New Orleans from Nature*. Baton Rouge: Louisiana State University Press.

Cutter, Susan. 2001. *American Hazardscapes: The Regionalization of Hazards and Disasters*. Washington, DC: Joseph Henry Press.

Downey, Liam. 2007. US Metropolitan-Area Variation in Environmental Inequality Outcomes. *Urban Studies* 44 (5–6): 953–977.

Elliott, James R., and Scott Frickel. 2011. Uncovering Environmental Consequences of Urban Change: The Production and Conversion of Relict Waste Sites in Portland and New Orleans. *Journal of Urban Affairs* 33 (1): 61–82.

Freudenburg, William R., and Robert Gramling. 1994. *Oil in Troubled Waters: Perceptions, Politics, and the Battle over Offshore Drilling*. Albany: State University of New York Press.

Frickel, Scott. 2008. On Missing New Orleans: Lost Knowledge and Knowledge Gaps in an Urban Hazardscape. *Environmental History* 13 (4): 643–650.

Frickel, Scott, and M. Bess Vincent. 2010. Katrina's Contamination: Regulatory Knowledge Gaps in the Making and Unmaking of Environmental Contention. In *Dynamics of Disaster: Lessons in Risk, Response, and Recovery*, ed. Rachel A. Dowty and Barbara L. Allen, 11–28. London: Earthscan.

Frickel, Scott, Richard Campanella, and M. Bess Vincent. 2009. Mapping Knowledge Investments in the Aftermath of Hurricane Katrina: A New Approach for Assessing Regulatory Agency Responses to Environmental Disaster. *Environmental Science and Policy* 12 (2): 119–133.

Frickel, Scott, and Michelle Edwards. 2010. Untangling Ignorance in Environmental Risk Assessment. Paper presented at the International Conference on Carcinogens, Mutagens, and Reproductive Toxicants: The Politics of Limit Values and Low Doses in the Twentieth and Twenty-First Centuries, Strasbourg, France, March 30.

Frickel, Scott, and James R. Elliott. 2008. Tracking Industrial Land Use Conversions: A New Approach for Studying Relict Waste and Urban Development. *Organization and Environment* 21 (2): 128–147.

Galison, Peter. 2009. Removing Knowledge: The Logic of Modern Censorship. In *Agnotology: The Making and Unmaking of Ignorance*, ed. Robert N. Proctor and Londa Schiebinger, 37–54. Stanford, CA: Stanford University Press.

Gross, Matthias. 2010. *Ignorance and Surprise: Science, Society, and Ecological Design*. Cambridge, MA: MIT Press.Hess, David. 2007. *Alternative Pathways in Science and Industry: Activism, Innovation, and the Environment in an Era of Globalization*. Cambridge, MA: MIT Press.

Hird, John A. 1994. *Superfund: The Political Economy of Risk*. Baltimore: Johns Hopkins University Press.

Latour, Bruno, and Steve Woolgar. 1986. *Laboratory Life: The Construction of Scientific Facts*. Princeton, NJ: Princeton University Press.

Leigh, Nancy Green, and Sarah L. Coffin. 2000. How Many Brownfields Are There? Building an Industrial Legacy Database. *Journal of Urban Technology* 7 (3): 1–18.

Logan, William Bryant. 1995. *Dirt: The Ecstatic Skin of the Earth*. New York: W. W. Norton.

Markowitz, Gerald, and David Rosner. 2002. *Deceit and Denial: The Deadly Politics of Industrial Pollution*. Berkeley: University of California Press.

Michaels, David, and Celeste Monforton. 2005. Manufacturing Uncertainty: Contested Science and Protection of the Public's Health and Environment. *American Journal of Public Health* 95 (no. S1): S39–S48.

Mielke, Howard W., Christoper R. Gonzales, M. K. Smith, Paul W. Mielke. 1999. The Urban Environment and Children's Health: Soils as an Integrator of Lead, Zinc, Cadmium in New Orleans, Louisiana, U.S.A. *Environmental Research* 81:117–129.

Mielke, Howard W., Eric T. Powell, Christopher R. Gonzales, Paul W. Mielke Jr., Rolf Tore Otteson, and Marianne Langedal. 2006. New Orleans Soil Lead (Pb) Cleanup Using Mississippi River Alluvium: Need, Feasibility, and Cost. *Environmental Science and Technology* 40 (8): 2784–2789.

Montgomery, David R. 2007. *Dirt: The Erosion of Civilizations*. Berkeley: University of California Press.

Nash, Linda. 2007. *Inescapable Ecologies: A History of Environment, Disease, and Knowledge*. Berkeley: University of California Press.

Noonan, Frank, and Charles A. Vidich. 1992. Decision Analysis for Utilizing Hazardous Waste Site Assessments in Real Estate Acquisition. *Risk Analysis* 12 (2): 245–251.

O'Neil, Sarah G. 2007. Superfund: Evaluating the Impact of Executive Order 12898. *Environmental Health Perspectives* 115 (7): 1087–1093.

Pellow, David N. 2007. *Resisting Global Toxics: Transnational Movements for Environmental Justice.* Cambridge, MA: MIT Press.

Proctor, Robert N., and Londa Schiebinger, eds. 2008. *Agnotology: The Making and Unmaking of Ignorance.* Stanford, CA: Stanford University Press.

Saha, Robin, and Paul Mohai. 2005. Historical Context and Hazardous Waste Facility Siting: Understanding Temporal Patterns in Michigan. *Social Problems* 52 (4): 618–648.

Schiebinger, Londa. 2009. West Indian Abortifacients and the Making of Ignorance. In *Agnotology: The Making and Unmaking of Ignorance*, 149–162. Stanford, CA: Stanford University Press.

Shapin, Steven, and Simon Schaffer. 1985. *Leviathan and the Air Pump: Hobbes, Boyle, and the Experimental Life.* Princeton, NJ: Princeton University Press.

Smith, Jodie, and James Rowland. 2007. Temporal Analysis of Floodwater Volumes in New Orleans after Hurricane Katrina. In *Science and the Storm: The USGS Response to the Hurricanes of 2005*, 57–61. Reston, VA: US Geological Survey.

Smithson, Michael. 1990. Ignorance and Disaster. *International Journal of Mass Emergencies and Disasters* 8(3):207–235.

Solomon, Gina M., and Miriam Rotkin-Ellman. 2006. *Contaminants in New Orleans Sediment: An Analysis of EPA Data* (February). Natural Resources Defense Council, Washington, D.C.

Tarr, Joel A. 1996. *The Search for the Ultimate Sink: Urban Pollution in Historical Perspective.* Akron, OH: University of Akron Press.

Timmons, Robert J., and Melissa M. Toffolon-Weiss. 2001. *Chronicles from the Environmental Justice Frontline.* New York: Cambridge University Press.

Tuana, Nancy, and Shannon Sullivan, eds. 2007. *Race and Epistemologies of Ignorance.* Albany: State University of New York Press.

US Environmental Protection Agency. 2006. Summary Results of Sediment Sampling Conducted by the Environmental Protection Agency in Response to Hurricanes Katrina and Rita. http://www.epa.gov/katrina/testresults/sediments/summary.html (Accessed February 6, 2012).

US Government Accounting Office. 2005. Chemical Regulation: Options Exist to Improve EPA's Ability to Assess Health Risks and Manage Its Chemical Review Program. http://www.gao.gov/new.items/d05458.pdf (Accessed February 6, 2012).

5

What Gets Buried in a Small Town: Toxic E-Waste and Democratic Frictions in the Crossroads of the United States

Phaedra C. Pezzullo

Rubbing two sticks together produces heat and light; one stick alone is just a stick. As a metaphorical image, friction reminds us that heterogeneous and unequal encounters can lead to new arrangements of culture and power.
—Anna Lowenhaupt Tsing

Nothing is ever truly buried, especially not garbage. The ritual of toxic waste burial serves only as a transitional phase—one that rears its ugly head periodically as cleanup technologies fail, contaminated bodies show signs of illness, and impacted communities begin to ask related public health questions. Remembering what we have buried, therefore, remains a critical act for rebuilding more democratically just and environmentally sustainable relations. To help foster such critical memory work, this chapter explores the historical specificity of one community's garbage culture as it relates to one toxin. By foregrounding culture in this account, I aim to mobilize Tsing's understanding of culture as friction: "the awkward, unequal, unstable, and creative qualities of interconnection across difference." (2005, 4). This history will identify key moments of friction, in which a variety of roles played by representative governments, residents, workers, and polluting industries negotiate power and culture. The heterogeneous results of these crucial instances of friction reveal some predominant myths and overlooked facts about the interconnectedness of the public and private spheres as well as the rural and technological realms.

The community of interest to this chapter is the place I now call "home." Other than serving as the home of "small-town" champion and rock superstar John Mellencamp and the flagship campus of Indiana University, the US city of Bloomington, Indiana, probably is best known in the national imagination as the location of the 1979 Academy Award–winning film *Breaking Away* (1979).[1] Dramatizing the common pattern of a town-gown divide, the film depicts two distinct ways of living in a small US city. That rivalry is significant, but it obscures another historical pattern about class and mobility: the town-company divide. That is, there is also a profound cultural divide in US cities like Bloomington between the corporations that build plants and provide jobs, on the one hand, and the residents and workers who live near the plants and/or are employed by them, on the other hand. Such histories both involve and exceed what environmental justice advocates call "job blackmail," where people are asked to choose between a healthy life and a way to make a living.

Though e-waste is a term now becoming popularized for the astounding numbers of computers and mobile communication technologies that are disposed of daily, the garbage dilemmas of e-waste have existed since the founding of electronic media technologies. From the 1950s through the 1970s, a major source of employment in Bloomington and the surrounding county of Monroe was provided by electronic industries. Westinghouse Electric Corporation located a factory in Bloomington during this time to manufacture electric capacitors; for insulating material, it used a mixture called Inerteen, which was a commercial name for toxins called polychlorinated biphenyls (PCBs) (Lane 1990a). In a three-year ban initiated in 1976, the Toxic Substances Control Act outlawed the commercial manufacture, sale, and distribution of PCBs in the United States.[2] For over two decades prior to that, however, corporations produced these toxins in unsafe working conditions and dumped their e-waste throughout the area.[3] In Bloomington, a complicated history of toxic e-waste pollution and democratic friction continues to unfold as a result of these practices.

This chapter provides a history spanning over seven decades of PCB waste in Bloomington. I have lived in Bloomington since 2002, have written public comments for and about the US Environmental Protection Agency (EPA) as a commissioner on the City of Bloomington's Environmental Commission about the community's right to know, and have experience supporting community-based efforts to clean up PCBs elsewhere (Nolan 2009; Pezzullo 2001). Here, I offer an overview of Bloomington's PCB history from 1930 to 2008.[4] For the sake of clarity, I have organized this story into two main periods: the invention, production, and distribution of PCBs (1929–1970), and the democratic frictions created between everyday people exposed to PCBs and the corporations they wished to hold accountable (1971–2008). The second period is organized into two additional parts, one focusing on tactics of residents who did not work at the plant, and the other focusing on tactics of workers who did work at Westinghouse's Bloomington plant.

This tale of rurality and pollution in some ways is indicative of broader geographic trends of US histories of waste. What is notable is not a rigid binary between rural and urban but rather how a research interest in what Tsing (1993, 288) calls "out-of-the-way-places" requires that we foreground marginality to some degree. Within the United States, despite advertising promising that we are all connected, we still live apart in many ways, geographically, economically, and culturally. As Valerie L. Kuletz (1998, 14) notes in her research of nuclear waste in the deserts of the US "wasteland," the pattern of exploiting regions for their natural resources "renders not only the land but the people who live on it expendable." Likewise, in his research of environmental injustices, R. R. Higgins (1994, 252) observes that communities historically segregated from elite centers of power are deemed "appropriately polluted spaces." When certain populations and waste are linked together, in what Mary Douglas (1968, 40) calls "separate areas of existence" and what Robert D. Bullard (1993, 12) names "sacrifice zones," US public culture can more readily forget the costs of toxic pollution because they—the waste and people disproportionately affected by the waste—appear hidden.[5] The labor of uncovering these stories and publicizing these struggles is thus counterhegemonic unto itself. While some cultural critics and students may find the technical and local specificity of any given story mundane, these histories remain significant to recover and retell if we wish to challenge dominant patterns of marginalization.

Bloomington's PCB history also provides us with an occasion to revisit conceptual assumptions about US culture, rurality, and technology as well as how people who work and live in these spaces are not just conceptually but also literally transformed by toxics. Paying heed to this tale may reveal some of the ways the dustbin of history is becoming corroded, and leaking into places and bodies living in the present.

The Birth of a Toxin, 1929–1970

Although the majority of the US population no longer lives in rural America, and cities remain more central to our popular imaginary, rural America stubbornly continues to be a significant and indispensable facet of US life. Urban America needs rural America for more than mythological roots perpetuated by pop songs and films. Bloomington, for instance, provides agriculture for food, limestone for building, and electronic technologies for, among other things, amusement and communication. As such, bringing small towns into our discussions about garbage can help us reconceptualize rural and urban boundaries, making us more account-able to the ways that public culture is, at least in part, constituted by moments of friction between the two, rather than as a one-way trajectory in which natural resources are mined and produced in rural areas, and then consumed in urban ones. As opposed to imagining such relations as unimpeded "flows," Tsing (2005, 5–6) reminds us that "friction inflects historical trajectories, enabling, excluding, and particularizing."

An example of the friction between rural and urban areas can be witnessed in the US history of PCBs. PCBs were invented as a way to reuse the industrial waste that is produced as a result of technologies that enable travel and commercial links between rural and urban places. At the start of the twentieth century, with the creation of assembly-line-produced automobiles, the demand for petroleum rose. When gasoline is extracted from crude oil, leftover chemicals or "by-products" are made as well. To discover if any of these by-products could be reused, companies began funding research. This initiative led to the invention of PCBs, which are the result of combining benzene, chlorine, and several other components (209 in total) (Montague 1993). In 1929, PCBs were manufactured by the Swan Corporation, which later became part of Monsanto Chemical Company, which then licensed the product to other companies under trade names such as Aroclor and the aforementioned Inerteen (Risebrough and Brodine 1971, 243). PCBs are stable, insoluble compounds that conduct electricity but not heat, and could be used as insulators in electric transformers and capacitors as well as for other purposes (Montague 1989). When conceived, PCBs were considered an exciting invention. Monsanto "produced more than a billion pounds of PCBs in the United States" in the twentieth century (Phillips et al. 1989, 351).

During this period of popularity, "scientific knowledge about the dangers of PCBs has advanced along two tracks, one private and one public" (Francis 1994). As early as 1936, skin disease and death due to severe liver failure occurred among workers at the Halowax Corporation in New York City; Halowax funded a study in 1937 that confirmed the systemic impacts of PCB exposure. Another 1938 study documented similar results with PCBs manufactured by Westinghouse and General Electric (GE). Monsanto also had a 1947 scientific finding that it should "give warning" because "the toxicity of those compounds has been repeatedly demonstrated." Yet none of these privately funded studies were publicized, and workers continued to be reassured that there was no risk of working with this compound. For example, although General Electric accumulated forty-three references about the health risks of PCBs by at least 1956, a 1950

General Electric instruction manual for workers assured them that it "may be handled in the same manner as mineral oil" (quoted in Higgs 1990).[6]

Since only the virtues of PCBs were touted publicly at the plants and in the communities where they were being produced, distributed, and used, the compound was not perceived as a threat. Bloomington was no exception. In the early twentieth century, Bloomington exemplified middle America. According to the 1910 US Census, Bloomington literally was the center of the US population. From 1890 to 1940, the center remained in the state of Indiana, until changes in population shifted the mean southwest (Winkle n.d.). Bloomington remained quite small in 1950. According to the 1950 US Census, the total population in dwelling units was only 21,021 people. The vast majority of those people were "white," with less than 200 "negroes" counted, and no other ethnicities were registered (US Department of Commerce 1953, 14–18). Although Bloomington was not large enough for the US Census to record income levels, the median income in 1949 for the state of Indiana was $2,116 per year (US Department of Commerce 1952, 14). Thus, despite the fact that Bloomington is not racially or ethnically reflective of broader polluting patterns shaped by environmental racism, it is economically and geographically reflective of historical patterns of environmental classism and rural marginalization.

Bloomington reflected the times in the twentieth century, growing to include improved infrastructure, transportation, and medical facilities. Limestone is Bloomington's best-known industry, providing materials for homes throughout the Midwest as well as urban iconic US structures such as the Empire State Building, the Pentagon, and the more recent 2009 Yankee Stadium. Responding to demands of the Industrial Revolution, the electronic industry also grew in Bloomington during the twentieth century. In the 1940s in Bloomington, RCA began producing radios and then, after World War II, televisions. Notably, in March 1954, "the first mass-produced color television set in America was manufactured at the RCA plant in Bloomington" (Anderson 2005).

General Electric and Westinghouse Electric Corporation soon followed by building plants of their own (Monroe County History Center n.d.). Founded in 1886 as the Westinghouse Electric Company, the Westinghouse Electric Corporation (Westinghouse) opened a plant in Bloomington in 1957. During its operation, it trucked approximately thirty-nine million pounds of PCBs from Monsanto into Bloomington. In 1989, Westinghouse sold the Bloomington plant to Asea Brown Boveri (ABB), a Swiss-based company, which kept it open until 2000.[7]

Although the plant did not open until 1957, the groundbreaking ceremony for the Westinghouse Bloomington plant occurred on Saturday, September 15, 1956, as part of the Monroe County Fall Festival. The 277,000-square-foot plant was built on a 113-acre tract three miles northwest of the city, and boasted of "the most modern and efficient equipment attainable." No mention of PCBs or Inerteen was made in the local newspaper. Westinghouse did promise that although technicians and engineers from out of town would fill approximately 100 jobs at the plant, 350 additional jobs would be available for local people and would involve on-the-job training (Top Westinghouse Men 1956; Groundbreaking Ceremony 1956). The jobs were warmly welcomed: "Bloomington has looked forward eagerly to the start of production to ease what has been a critical shortage of jobs for men in heavy industry. The Bloomington Advancement Association has said the arrival of Westinghouse will greatly ease the situation" (Steel for W.E. Plant 1957). For over a decade, Westinghouse seemed like any other large corporation that provided much-needed jobs and skills training. Slowly, however, the company's toll on the community began to show as well.

"New Symbols of Possibility," 1971–2008

By 1971, PCBs were detected everywhere, in everything from Arctic birds to the breast milk of women in California. Through various global studies, a scientific consensus quickly emerged that PCBs were ubiquitous, persistent, and dangerous to human health as well as broader ecosystems (Risebrough and Brodine 1971). According to the Agency for Toxic Substances and Disease Registry (2001), PCBs have been linked to the following health complications: skin conditions (such as acne and rashes); liver, stomach, and thyroid gland damage; anemia; compromised immune systems; behavioral alterations (such as problems with motor skills and a decrease in short-term memory); impaired reproduction; and cancer. More recent studies have indicated that people who consume PCB-tainted food can pass on the chemical for six generations (Bovee 1988). Growing scientific evidence of the risks of PCBs led to a change in federal policy, which in turn led to renegotiations of culture and power throughout the United States, including in Bloomington.

In 1976, Russel E. Train, then EPA chief, launched a national campaign to eliminate PCBs. This federal effort was galvanized by a series of events. Notably, at an EPA conference on PCB dangers in Chicago the year before, the Bloomington city chemist attended and returned home to start local investigations. High levels of PCBs were detected in the Bloomington sewer system and the fish downstream from the Westinghouse Bloomington plant, which led to the warning not to eat locally caught fish and additional investigations (Schroeder 1991). Contamination from the Westinghouse plant occurred through use on-site and the local dumping of PCB-laced materials, but also through the free distribution of PCB-laced sludge to farmers and gardeners for fertilizer, which was "the primary means of disposing of sludge in the Bloomington area" (Spreading Stain 1976). Once in the broader ecosystem, PCBs spread into the food supply through fish, crops, the sewage system, and livestock. One family drank milk from a cow that was found to have been exposed to double the maximum PCBs levels allowed by law (Livezey 1980). Studies found that Bloomington had "the nation's largest volume of PCBs—650,000 cubic yards of landfill soil" (Schroeder 1991). As I elaborate below, occupational exposure for some people was also extreme.

When asked why 150 residents in Bloomington were being given blood tests for PCBs in 1984, Greg Steele, a chronic disease epidemiologist with the State Board of Health, claimed the ubiquitous exposure has had a profound impact on community health:

Bloomington is an excellent community to study because the residents have been exposed to so many routes. Those who will be tested include people who have eaten contaminated fish, tainted game, have scavenged metal from capacitors, played in or around dumps containing PCBs, swam in contaminated quarries or were exposed occupationally. (quoted in Test for PCB Contamination 1984)

Notably, Monroe County also has what is called "karst topography," in which limestone quarries have been bountiful due to sinkholes, fissures, and underground streams. It was common practice before regulations changed for companies and residents to dump waste in these sinkholes as well as along roads, in woods, or into waterways. The karst topography makes PCB cleanup more challenging because there is no way to follow all the intricate underground paths that this toxic waste has followed. In this sense, the geographic topography provides a subterranean mystery to this history, much like the ways that tracing the sources of and interactions between toxins in human bodies often continue to test science's capacity to establish cause-and-effect relationships. Undetectable by the naked human eye, PCBs travel routes inside and outside our bodies by taking hold in spaces that our imaginations still cannot follow with ease.

Bloomington's PCB history follows a pattern that Michael R. Reich has identified after studying three toxic global disasters. According to his research, communities that have experienced toxic disasters tend to move their understandings of environmental pollutants from a nonissue (that is, when no one believes there is anything wrong) to a public issue (when people begin to share stories and express concern) to a political issue (when a collective forms to gain redress). Reich (1991) also argues that while communities that have experienced toxic disasters attempt to publicize the surrounding controversy following this pattern of increased social awareness and mobilization, institutions that have been responsible for the pollution try to privatize the surrounding controversy to limit awareness and accountability. The pages below note these patterns of democratic friction involving residents and then workers.

Residents' Struggles

To recover the costs of cleanup already incurred and predicted for the future, the City of Bloomington pursued Monsanto in court for seven years, from 1981 until 1988. During this time, Monsanto made two primary arguments. First, it claimed that the benefits of PCBs outweighed the risks, which it downplayed. As Michael Fruehwald, an attorney representing Monsanto alleged: "There's no question there are toxic effects to PCBs. But the benefits of PCBs were so great, and the toxic effects of PCBs were so controlled that it . . . would have been irresponsible for Monsanto to stop selling PCBs before it did" (quoted in Hinnefeld 1988e). The second argument followed one that was successfully used by contemporary gun manufacturers—namely, that the production of guns is not the cause of gun violence; the people who purchase guns are.[8] The court agreed that in the case of PCBs, liability should not be found for "merely manufacturing dangerous products" but instead in the use; since it was Westinghouse that used the product, the courts absolved Monsanto of all liability for the environmental pollution from the Westinghouse Bloomington plant, claiming it was "unwilling to extend the doctrine of strict liability for an abnormally dangerous activity to the party whose activity did not cause the injury."[9]

"Friction," as Tsing (2005, 6) points out, "is not a synonym for resistance. Hegemony is made as well as unmade with friction." Noting the roles of dominant cultural institutions in this process, Reich (1991, 235–251) has found that polluting institutions tend to privatize toxic pollution controversies through strategies of dissociation, confrontation, and diversion. Whereas Monsanto successfully dissociated itself from polluting the area and diverted attention to all the good it provided to the community, Westinghouse was forced to confront the federal and local governments' calls for accountability, and so was unsuccessful in privatizing. In 1983, the US Department of Justice initiated a civil action against Westinghouse over two of the sites most contaminated with PCBs, Neal's Landfill and Neal's Dump, under the Comprehensive Environmental Response, Compensation, and Liability Act (often called "Superfund"). Then the City of Bloomington sued over two additional sites. The courts consolidated these two cases and added two more sites for a total of six. In 1985, a legal settlement called a "consent decree" was reached between the state of Indiana, Monroe County, and the City of Bloomington with Westinghouse in order to clean up these six sites of PCB contamination, costing between seventy-five and one hundred million dollars. At the time, EPA Administrator Lee M. Thomas announced: "Resolution of this enforcement case represents the largest hazardous waste settlement in the history of the agency. It is a comprehensive agreement, which provides for the ultimate destruction of the PCB wastes, rather than for long-term landfilling." The plan was for Westinghouse to construct an incinerator south of the city, and over fifteen years, remove, transport, and incinerate the 650,000 cubic yards of contaminated waste.[10]

Many expressed concern that the six sites did not adequately cover the contamination. Eventually, the EPA studied 189 suspected PCB sites to potentially be added to the Comprehensive Environmental Response, Compensation, and Liability Inventory System list. "They include about 130 properties where PCB-laden sewage sludge was used on gardens and about 60 old dumps and yards where PCB-filled capacitors from the Bloomington Westinghouse plant were broken open for salvaging [for copper]."[11] Part of the challenge of testing further, however, was that private property owners needed to consent to it, and many did not want the stigma and financial devaluation of their property if they were officially declared a PCB-contaminated site (Hinnefeld 1988h).[12]

After the announcement of the consent decree, many local residents also raised concerns about inadequate health monitoring and cleanup as well as how Westinghouse could profit from the incinerator contract (Trewhitt and Rich 1985). Some expressed mistrust of Westinghouse. Resident Mike Baker, for example, argued: "Westinghouse has polluted Bloomington with more PCBs than any place in the country. I really don't believe we can trust them to build an experimental incinerator" (quoted in Sheckler 1990). This distrust of polluting corporations is common for the most severely impacted communities, and is part of why appeals to public decision-making processes, judicial forums, and elected representatives become salient. When communities cannot trust the "experts" (who

often calculated that the benefits to residential/worker lives could out-
weigh the risks), insisting on more democratic practices becomes a means
to account for costs beyond the economic bottom line.

In addition to concerns that a Westinghouse-run incinerator therefore
was not the healthiest or safest solution for the community, many residents were infuriated that their voices were left out of what they believed
should have been a more democratic decision-making process. It was not
until 1988 that Bloomington residents (including attorney David Schalk
and city chemist Ron Smith) successfully sued the federal government
to obtain copies of the federal court documents that led to the consent
decree agreement.[13] Monroe County also endorsed hiring a Boston-based
company to consult with the local board of health on the incinerator
permit (Higgs 1988a). When the City of Bloomington hired a Washington, DC, firm to assess the incinerator, they found that Westinghouse had
failed to consider all the possible public health risks (for instance, it ignored the impact on local farming families that raise their own food) as
well as the quantity of PCBs in the emissions that may be produced (Hinnefeld 1988b). This led the city to sue Westinghouse.[14] The plans to build
the incinerator were delayed as a result of this public opposition, showing
the local community that in this case, their voices could sway legal and
political representatives.

Democratic resistance from local citizens continued throughout the
next decade in Bloomington in response to the consent decree. Although
many of the landmark decisions were settled in the courts, public pressure on political officials to pursue these cases occurred during public
hearings, in grassroots meetings, and through marches (some of which
resulted in arrests).[15] In 1976, Jon Canada and others formed the first
related grassroots group called Citizens Concerned about PCBs; it dissolved relatively quickly, though, once residents believed the government
was addressing the problem (Sheckler 1991a). The proposed incinerator
changed that impression, mobilizing the grassroots community rapidly.
A petition drive against the incinerator resulted in more names than had
voted for Bloomington's mayor in 1983 (Andrews 1989). By 1984, Canada, Schalk, Mick Harrison, Marc Haggerty, Mike Andrews, and about
ten others formed the Toxic Waste Information Network to nationally
publicize their opposition to the proposed incinerator.

In 1987, resident Margo Blackwell cofounded PATI, People Against
The Incinerator. One of their most significant events took place on Sun-
day, April 16, 1989, when Blackwell organized a march with Greenpeace to raise awareness about their concerns. Over five hundred people

participated (Beaven 1989).[16] In 1990, Greenpeace highlighted Blackwell and other Bloomington activists from this PATI event in a VH-1 public service announcement (Creek 1990). With narration from actor Alec Baldwin, the two-minute segment aimed to celebrate the twentieth anniversary of Earth Day by encouraging citizens to become more involved in democratic environmental struggles. The tagline was "ORDINARY PEOPLE ARE DOING EXTRAORDINARY THINGS." Blackwell reinforced this message: "I thought it was time that somebody should basically take the bull by the horns and say, 'No, this is not right, and we're not going to let you do it.'" Fellow resident and activist Patti Cummings did the same, emphasizing that "you can't wait for anybody else to do it. You have to do it yourself. And if you don't do it yourself, it very likely will not get done (VH-1 WORLD ALERT 1990)"[17]

One common response by institutions under scrutiny for toxic pollution is to discredit the victims (Reich 1991, 241).[18] For example, some stereotyped early protesters in Bloomington as "radical hippies"—a rhetorical move that attempts to marginalize and trivialize the critiques being performed. This strategy did not always work, as one county commissioner remarked to another at the time: "What he is saying is right, even if his hair is too long." To continue to mobilize resistance, residents picketed, sang protest songs, and set up a display of PCB-soaked items at the local library (which caused the shut down of the library for a week for detoxification). As David McCrea, a local attorney, commented about his decision to bring PCB-contaminated materials to the mayor's office, he wanted to make the contaminants more present to those who didn't live on the west side of town: "I didn't do it like a prank. I wanted her to see what one looked like and smelled like. I wanted to impress upon her that there was a problem that anyone could see and that should be cleaned up" (quoted in Sheckler 1991a).[18]

Starting in 1989, this more marginal and radical group grew once news spread not only of a proposed incinerator but also of a local landfill for the incinerator ash. As Westinghouse spokesperson Kit Newton recalled:

I think there was sort of a dull roar of opposition before that, but once we announced the landfill (on Dec. 6, 1990), I think the project became real to more people. I think the opposition today has more credibility than it did before. People are sincere and have legitimate concerns. And I have no doubts whatsoever that this will be the most closely watched and regulated incinerator in the United States. (quoted in Sheckler 1991a)

The landfill led to the formation of the Coalition Opposed to PCB Ash in Monroe County (COPA), led by President Mike Baker. In addition to organizing public forums, the group produced a newsletter (which had a mailing list of two thousand at one point), television ads, T-shirts, stickers, and buttons. Blackwell commented: "I say 'Thank God for COPA.' Some mainstream people have been behind the opposition for a long long time, but COPA has made it acceptable to be against the incinerator." One resident and vice president of COPA, Jim Shea, announced: "I've been labeled an anti-incinerator activist but I don't feel like an activist. I just feel like an ordinary citizen who has looked at this plan and found it doesn't make any sense" (quoted in Sheckler 1990).

It took Westinghouse four years to propose an actual plan to build an incinerator (Creek 1989a). During the subsequent sixty-day comment period, public opposition continued. On November 21, 1989, the EPA inaugurated a Bloomington Community Advisory Committee to enable citizens to have a greater voice in the decision-making process (Creek 1989b).

In 1991, the Indiana state legislature introduced a bill that then-governor Evan Bayh signed into law requiring the Indiana Department of Environmental Management to seek a safer alternative to incineration and the local county to include such a provision in its solid waste management plan (Hinnefeld 1991; Van der Dussen 1991). In 1994, after almost a decade of intense grassroots and political pressure, Westinghouse finally put its plans to build an incinerator on hold as it sought alternatives (Sheckler 1994a, 1994b).

Eventually, Westinghouse, the EPA, and local officials agreed on a combination of on-site remediation and shipping contaminated soils to other states. Westinghouse began cleanup at the end of August 1997, over twenty years after the PCB contamination was initially discovered (Hinnefeld 1997). This site-by-site cleanup process continues today, as does some residents' skepticism as to whether the process is being completed to the highest level of decontamination that is technologically feasible or those in charge of the cleanup are open to the concerns of citizens' voices.[19]

Workers' Struggles

Even when workers are residents of a community, the rhetorical constraints on their ability to mobilize democratic decision-making processes differ. On the one hand, workers frequently have increased knowledge of what is happening at an industrial facility, particularly in terms of the names of specific toxins involved, their uses, and the ways they are disposed. On the other hand, worker frictions pose the risk of larger conflagrations, because they often face greater pressure to stay silent about concerns unless they want to risk losing their jobs. Friction, in this context, can be harder to engage constructively, even if the impacts are corporeally more apparent.

Reflecting national trends, even when workers at the Westinghouse Bloomington plant asked questions, they were adamantly reassured that PCBs posed no risk to their health. For example, in 1968, when news that thirteen hundred residents of Kyushu, Japan, had become ill with lesions, ailments, birth defects, and other public health problems from eating PCB-contaminated rice oil made global headlines, Bloomington workers wondered if there was a connection between those exposures and their own. To ease potential tensions, the plant manager at the time, Donald M. Sauter, "dipped his hands" into PCBs at an employee meeting to persuade workers that there was no reason to be concerned. While this bravado might have been the public face of the company, Westinghouse followed the historical pattern of privatizing potential conflicts while funding a 1972 study that identified massive contamination in the Bloomington area; the company, unfortunately, never shared the results with workers, residents, or the local government (Schroeder 1991).[20]

When national events and local controversy eventually exposed this corporate culture of secrecy, Westinghouse discouraged workers from joining the local environmentalists and anti-PCB protesters through what environmental justice advocates call "economic blackmail" (Bullard 1993, 12–13, 23).[21] According to one employee, Mont Toon, "They would tell the people in the plant that the 'radicals' were trying to run us out of town and that we'd all be out of jobs. They told us that if we didn't go out and talk at town meetings about how it wouldn't hurt you, we'd lose our jobs" (quoted in Lane 1999a).

Workers' attitudes about PCBs usually depended on their exposure and health. Of the approximately 3,588 people who worked at the Bloomington plant from 1957 to 1977, some say they never were exposed and tend to think the issue has been overblown. Some did not take the concerns seriously until the latent exposure impacted them. Others have spent decades struggling with health issues related to their direct exposure through skin absorption and inhalation (Lane 1999a; Sheckler 1992a).

Ralph Evans's job, for instance, began after other workers had gone home, when he washed the remaining Inerteen down the floor drains of the plant (Lane 1999a). After working at the plant from 1965 to 1981, he retired on medical disability at thirty-eight years old. Evans has suffered a variety of health problems, including a bone disease, a blistered esophagus, tumors on his tonsils and gallbladder, adhesions on his intestines, brain lesions, and more. For decades, he was frustrated by Westinghouse's denial of responsibility: "As long as Westinghouse says it won't hurt you, no one will ever get any help. That burns me up, but what can you do? They don't want to hear about it and they don't want anything to do with you. We're left here to deal with it. There's no end to it" (Lane 1991a, 1990c, 1991b). [1]

Jason Morrow worked at the plant from 1965 to 1978, but made it only three months on the Inerteen line, where he worked directly with PCBs by filling capacitors with Inerteen and soldering them shut. At the time, he broke out in a rash on his exposed body parts (his hands, arms, and face) and said that the protective goggles were not adequate for his burning eyes. Morrow knew something was wrong when the fluid managed to eat through the laces on his boots: "I had combat boots from the service. They were leather boots and it would eat through the threads and they'd fall to pieces" (quoted in Lane 1990b). The boots were exposed because Inerteen spills were common, covering the floor. Some workers recall that the floors were covered with sawdust in an attempt to soak up the slick, greasy liquid. According to another worker, Gilbert "Gib" Prow, "A pair of shoes would last about three months. They'd get as hard as a bone and curl up at the toes" (quoted in Lane 1990a; see also Higgs 1990.). Once his complaints were heard, Morrow transferred to another section of the plant, but as a union representative, he continued to advocate for better protection for those on the Inerteen line. Transferring employees to a different part of the plant after health complaints became a pattern of institutional diversion and deflection regarding worker concerns (Lane 1990).

Rick Sluder, former chief union steward, recalled perhaps the most extreme exposure and related health impacts:

I used to come home from work every night soaked with PCBs, saturated right down to my underwear. Now I think I'm paying for all those years when I worked on the Inerteen line at Westinghouse. . . . That Inerteen line was a nightmare. You'd get it in your eyes, your hair and ears and nose and you could never wash the smell off you. The old PCB oil would stink up the dryers at the laundromat so bad that they asked me and my wife to stop coming there. . . . My feet would blister and bleed, my toes would swell like big pancakes. And my skin got so it wouldn't heal right. I had like big scars around my waist where my belt and pants were.

In addition to the immediate health complications that became apparent, Sluder had long-term health impacts, including impaired memory, inflammation and pain throughout his body, a tumor in his right forearm, and headaches—all by the age of forty (Sheckler 1992; Schroeder 1991).

Sluder found out that he had PCBs in his blood after being part of a 1977 study conducted by the National Institute of Occupational Health and Safety of Westinghouse workers. He was told four years later that he had a PCB concentration of 3,450 parts per billion in his blood—the highest level of PCB contamination of a human recorded in the world. The only advice he was given on notification of the results was not to donate blood to others, and not to be buried in a regular cemetery because his body was so toxic that it would need to be placed in a hazardous waste landfill (Sheckler 1992b; Schroeder 1991). In this extreme case of legal dehumanization, Sluder's body was literally transformed by the standards of US law from that of a human being to toxic waste. Such a traumatic transformation seems notably lost for those who have not heard of Bloomington's PCB legacy.

Once seemingly undeniable risks began to surface through the bodies of workers, Westinghouse eventually did upgrade working conditions to make the job relatively safer. Morrow explained that "they made attempts to improve the job. They took precautions. They furnished boots, so that the Inerteen wouldn't eat your shoes off, and plastic coveralls and coats. Then they put in showers a few feet from the line to rinse off." Despite these precautions, Inerteen continued to be used (quoted in Lane 1990b). To discredit the victims of its hazardous practices once again, the company also was quick to point out other potential causes for health complications, such as smoking cigarettes or sun exposure (Lane 1991[22]

Although the courts found Monsanto not liable for polluting the area, workers from the Westinghouse plant have had somewhat more success in suing Monsanto for inadequate warning to those working with PCBs at the Westinghouse plant. In contrast to the environmentally based cases noted previously, workers focused on Monsanto instead of Westinghouse because federal law finds the manufacturer liable, not the purchaser.[23] Beginning in 1986, a national civil action was filed with fifty workers from Bloomington, Muncie (Indiana), and Cincinnati (Ohio), but the court separated the cases. The following people from the Bloomington plant were involved: Albert Fritch, Ken Siniard, David Wampler, Cleo Rader, James McElroy, Glenn Brown, and Gertrude Koon. Choosing not to focus on the cancer cases since they tend to be more controversial, lawyers for the Bloomington workers concentrated on a lack of adequate warning about the hazards and variety of other PCB-related health issues, including neurological disorders, arthritis, head pains, and rashes.[24] During the fourth week of the trial and just days before the jury was to decide, in November 1991, eight Bloomington workers won an out-of-court settlement with Monsanto, and the checks arrived in February 1992. Although part of the agreement was nondisclosure regarding the terms and Monsanto continued to claim this settlement was not a confession of liability, the workers involved considered it a democratic victory. According to Albert Fritch, "Winning against them was as much a moral type thing as anything. Just winning was enough satisfaction for me" (quoted in Higgs 1992b). David Baugh agreed that bringing legitimacy to public discourse about PCBs was the primary goal: "The monetary part of it wasn't the important part. I don't think anybody was in it for the money. It was bringing attention to the world about the effect of these chemicals" (quoted in Higgs 1992a).

In addition to some workers standing up for themselves as individuals at the plant and in the courts, the union continued to work to improve safety precautions, lobby for more health studies and accountability, and publicize education about the impacts of PCBs. As late as 1993, many workers remained afraid of being labeled an agitator; therefore, to increase democratic participation, the union began to request that public meetings about the cleanup process be videotaped and broadcast on the local cable station so workers could watch from home. According to Dan Potts, then president of the Plant Union, Local No. 2031 of the International Brotherhood of Electrical Workers, "We want people to be educated about PCBs. The average worker is sort of afraid of government and a lot of them won't go to meetings like this. They also feel like nobody cares about us so it's not any use" (quoted in Sheckler 1993).[25]

Despite some small victories for a handful of workers, concern about the risks of being associated with democratic resistance was understandable. Doctors with expertise in PCBs have never visited workers in Bloomington to explain the blood tests taken in 1977 or discuss how workers can mitigate the impacts, if at all. The National Institute for Occupational Safety and Health (NIOSH), Westinghouse, and Monsanto publicly continue to disavow a need for employees to be given special status. They argue that since PCBs have become ubiquitous, workers have not been necessarily more impacted, and further, that additional practices unrelated to work could be equally accountable. A study released in 1992 by Indiana state epidemiologist Gregory Steele did indeed find that the general population nationwide averaged between 7 to 10 parts per billion of PCBs, and workers fell within this range, with 7.8 for females and 8.1 for males. Those levels dramatically increased to 80 parts per billion for people who scavenged PCB capacitors or ate high quantities of local fish.

Yet Steele's study also revealed that people who worked in the production area at the Westinghouse plant had an average of 800 parts per billion, and those who worked in the oven area of the plant reached 1,800 parts per billion. Thus, while NIOSH, Westinghouse, and Monsanto seem to be partially correct in their stance (that is, that PCBs are ubiquitous, and everyday activities can change one's exposure), their continued denial that worker exposure is not necessarily exceptional seems unfounded (Sheckler 1992b).

Conclusion: The Rural Fact and Technological Myth

Some might say that the story of PCBs in the United States ended in 1976 when their commercial manufacture, sale, and distribution generally was outlawed. But what to do with the PCBs already created (and PCB use not covered by the Toxic Substances Control Act) remains a dilemma globally. The life cycle of a persistent organic pollutant like PCB is not a simple story of birth and death. "These poisons are now ubiquitous, and are especially concentrated in the flesh of predators. Potentially dangerous levels of PCBs can be found in the fatty tissues of seals, whales, eagles, many fish, and virtually every human on earth" (Francis 1994). Following national and international trends, the PCB pollution in Bloomington transformed from an invisible, innocuous substance into a visible, dangerous presence.

Although much of Bloomington's PCBs have been remediated and the city's demographics have changed, toxic e-waste has been transported and burned in incinerators located in other communities, and PCBs persistently circulate through and remain in the bodies of people exposed as well as the fish, birds, and other wildlife that live in the polluted waters that flow in and out of the complicated karst topography.[26] Monsanto and Westinghouse spent decades hiding studies they had conducted while reassuring the public that the benefits of PCBs outweighed any concerns. Once residents obtained news of the dangers of PCBs, a slow but steadily growing number demanded their right to know more information about their exposure and participate in the decision-making process about what to do with the e-waste. Mobilizing a democratic response to toxic waste has taken decades and a multipronged approach, including protests, hearings, court cases, legislation, and worker resistance. In instances of back-and-forth negotiations, friction is an apt metaphor—uneven, not always verbalized, but eventually creating burning questions that get addressed in productive, if not always desirable, ways.

The PCB story in Bloomington remains compelling in its specificity; yet, it also serves as a representative anecdote for the impact of modern industrial and chemical revolutions on the US landscape. Bloomington's role in the national production of limestone, electronics, and PCBs goads us to recognize that this story is not about some small town that has no bearing elsewhere. The circulation of resources, technologies, and pollution exceeds municipal boundaries along with culturally constructed borders between the rural and urban. Although the US population has shifted to urban centers, we must be careful not to dismiss rural life as something that no longer matters except in our social imaginaries. Doing so risks a fragmentary conception of our interconnectedness by forgetting the ways that urban US life is predicated on and intertwined with rural US life, and vice versa.

In addition, this history provokes us to reexamine the myth that technology will save us not only from the limits of rural life but also from the limits of technology itself. Leo Marx (1964, 226) once argued in his landmark *The Machine in the Garden* that the twentieth-century US condition was defined by the contradiction of "the rural myth and the technological fact." This observation referenced the significance of the pastoral ideal in the US public culture—born of a time when the majority of European colonizers lived in rural areas, but also of a long-gone culture, since the majority of the population has since moved to create urban centers. In his analysis of how the rural become an ideology, Marx turned to legendary US writers such as Henry David Thoreau, Herman Melville, Mark Twain, and Samuel Hopkins Adams to indicate the contradictory condition of America's belief in technological progress. Indeed, "no trace of untouched nature remains," and "until we confront the unalterable, . . . there can be no redemption from a system that makes men [*sic*] the tools of their tools" (ibid., 355). Marx contends that the symbols of possibility we need to challenge a condition of our own making must come from politics (ibid., 365).[27]

Yet even if Marx's abstract assertion was a useful diagnosis of a historical condition, the history of Bloomington's PCBs reflect the *fact* that the technological has continued to permeate more of our everyday lives than ever before, and in ways that aren't always anticipated or desirable. This story also reveals that US culture is indebted to the *myth* that technology will save us from technology. Afraid to admit limits to what should count as technological progress or unwilling to believe that humans are capable of creating technologies for which we have no cure, this chapter shows how the fantasy of technology's unambiguous promise played a pivotal role in Westinghouse and Monsanto perpetuating toxic pollution along with their reticence to clean it up. Moreover, as bodies that have experienced extreme exposure transform from the biological to the technological by current legal standards, the agency of humans to counter the technological myth seems all the more precarious. As such, through this account of Bloomington's history with PCBs, I hope to have demonstrated that the reverse of Marx's argument is true today: we must also be attentive to the rural fact and technological myth produced by various moments of democratic friction.

Bloomington, of course, has needed technology to remove and detoxify what PCBs could be located in its terrain, but remediation technology still cannot match the complexity of the karst topography or human body—each of which involves subterranean geographies that the best of our experts have yet to fully map or comprehend. Despite the importance of technology to the history and future of the United States, the myth of technology seems as significant as the fact. Further, as more people are exposed to greater amounts of toxics, the extraordinary tragedy of what happened to Sluder may become as ordinary as the broader inequitable patterns of e-waste, toxic pollution, and everyday people that Bloomington illustrates. Given this, we would do well to continue to investigate the ways certain places and people are marginalized as garbage, both metaphorically and materially.

Bloomington's PCB history is therefore revealing to those of us invested in studying frictions between goods, people, places, and toxics. The United States is long overdue for toxic legislation reform that can better account for the quantity and wide variety of toxics that have become a part of our everyday lives. In considering what we should do to bring about a more sustainable future, we would do well to learn from this twentieth-century tale of a small town's frictional history of toxic waste and resistance as well as continue to foster conversations on how we can better account for the spatial and cultural politics of the garbage we create and may be becoming.

Notes

1. In 1977, John Mellencamp bought a home just fifty-six miles away from his birthplace in Bloomington, where he has lived with his family and recorded since the mid-1980s.

2. As Fox River Watch (n.d.) notes, "Congress passed the Toxic Substances Control Act which outlawed the manufacture, sale, and distribution of PCBs except in 'totally enclosed' systems, within 3 years. It was the only chemical Congress

itself has ever banned. However, enclosed transformers and capacitors are STILL allowed to contain PCBs." In 1979, the Environmental Protection Agency (n.d.) "issued final regulations banning the manufacture of polychlorinated biphenyls (PCBs), after a 3-year phase-out period. In addition, the EPA rules gradually ended many industrial uses of PCBs over the next five years, but allowed their continued use in existing enclosed electrical equipment under controlled conditions."

3. Some famous instances include the illegal disposal and relocation of PCB-laced soil in North Carolina that galvanized a movement against "environmental racism" and "environmental injustice" (Pezzullo 2001) and the well-known dumping of over one million pounds of PCBs in the Hudson River by General Electric—a case that is still being debated in courtrooms at the time of this writing.

4. There is a resistance among many local residents who have lived in the area longer than I to share their personal histories about the PCBs in town, and what events have occurred in relation to their exposure and struggles. As a cause for their trepidation, many have noted the people who once were participants and have since died; others have suggested a fatigue with the saga, or with academics in a town with regular student and faculty turnover. My archival research began with the Web site hosted by the Coalition Opposed to PCB Ash in Monroe County, Indiana (http://www.copa.org), which includes extensive legal, scientific, and political documentation about PCBs, information on the local PCB Superfund sites, and perspectives from some of the key residents who have been involved in this ongoing drama. Then I visited the Monroe County Public Library, which is an official EPA document repository on PCBs in Bloomington. My university library provided me online access to national newspaper stories from 1930 until 1989, microfilm from the local newspaper, the *Daily Herald Telephone*, from 1956 to 1957, and copies of relevant legal cases. Through my online subscription to the local paper, now renamed the *Herald Times*, I also read 1,145 articles published from April 1, 1988 until the present day.

5. "If cleanness is a matter of place, we must approach it through order. Uncleanness or dirt is that which must not be included if a pattern is to be maintained. To recognize this is the first step towards insight into pollution. . . . [It] involves no special distinction between primitives and moderns: we are all subjects to the same rules. But in the primitive culture the rule of patterning works with greater force and more total comprehensiveness. With the moderns it applies to disjointed, *separate areas of existence*" (Douglas 1968, 40; emphasis added).

6. In the June 1936 edition of the *American Journal of Public Health*, Dr. Louis Schwartz warned of the health effects that PCB exposure could cause, including skin lesions and systematic poisoning. Cited in Higgs 1990. See also Lane 1990a; Francis 1994.

7. See Creek 1989c; Lane 1990a; Hinnefeld 2006; Sheckler 1994c. While Westinghouse will remain the main referent throughout this chapter for the company since it was the owner during the Bloomington plant's operations, due to the complicated dynamics of corporate conglomerates in the second half of the twentieth century, the liable company has shifted from Westinghouse to Viacom to the CBS Corporation. In 1995, Westinghouse bought CBS Corporation, but by 1997, it

renamed itself the CBS Corporation with Westinghouse as one brand managed by the CBS media conglomerate. In 1999, Viacom, Inc., purchased CBS Corporation and then, in 2005, Viacom also renamed itself the CBS Corporation. For a more detailed time line of Westinghouse's history, see Westinghouse 2009.

8. *Hamilton v. Accu-tek*, 935 F. Supp. 1307, 1324 (E.D.N.Y. 1996).

9. Ibid, 121–122; See also Hinnefeld 1988c. All subsequent appeals of this ruling were denied.

10. Litke 1985; MacNeil 1985; Shabecoff 1985; Peterson 1985; Trewhitt and Rich 1985; Hinnefeld 1988g.

11. Hinnefeld 1988a, 1988f; Higgs 1988b.

12. In 2000, the EPA offered to temporarily relocate three families whose homes were adjacent to the PCB contamination site at the Lemon Lane Landfill. The Griffin and Pelfree families moved, but the Elliott family stayed. Jerry Pelfree wished more was offered: "I think they ought to be made to buy all this property around here. This whole neighborhood's on a blacklist from getting any loans or anything to build houses" (quoted in Hinnefeld 2000).

13. *Schalk v. Reilly*, 900 F.2d 1091, 1096–97 (7th Cir. 1990); see also PCB Records to Be Turned Over 1988.

14. *City of Bloomington, Indiana v. Westinghouse Elec. Corp.*, 891 F.2d 611 (7th Cir. 1989).

15. As one elected official noted, "We would have an incinerator today if it hadn't been for those people who continued bringing information forward, despite getting beat back time and time again" (quoted in Higgs 1994).

16. In the 1990s, another local grassroots group, Citizens Opposed to PCB Ash in Monroe County Inc., formed under Mike Baker's leadership to continue to mobilize opposition against the incinerator and proposed landfill for the incinerator's ash (Sheckler 1991c).

17. Although clips of the march were shown, the only other resident voice was Cumming's child, Bo, who expressed concerns about getting sick from contamination.

18. The west side of town historically has been more industrial and working class than downtown or the east side. As one city council candidate claimed when running for office, "I think the west side has been treated as a forgotten parent. I would say it is a stepchild but Bloomington came from the west side and this area needs to have a more powerful advocate in city government" (quoted in Sheckler 1991b).

19. See, for example, *Frey v. E.P.A.*, 2005 WL 767057 (7th Cir. [Ind.] April 6, 2005).

20. Residents later obtained copies of a Westinghouse report from 1971 documenting the company's knowledge of contamination in local landfills and creeks (Hinnefeld 1988i; Lane 1999a).

21. Economic blackmail forces workers to choose between having a job and having acceptable occupational health standards.

22. For a compelling analysis of the common international pattern of industry privatizing and deflecting in response to communities trying to publicize and hold accountable toxic polluters, see Reich 1991.

23. In 1987, three Bloomington workers—Medus Hutchens, Ralph Evans, and Mont Toon—went to trial in Texas federal court with ninety-seven workers from five states and lost. While the initial jury ruled in Monsanto's favor, the judge overturned the decision, only for a federal appeals court to reinstate the original ruling (Lane 1990c).

24. In 1989, the Indiana State Board of Health released a report on the increased cancer death rates among Westinghouse Bloomington plant workers. It found that men and women at the plant were twice as likely to die from brain cancer compared to rates for the general population. It also found white males working at the plant were twenty times as likely to die of malignant melanoma (Creek 1989c).

By 2006, the National Institute for Occupational Safety and Health (NIOSH) also confirmed PCB exposure increases one's risk of contracting cancer. For a link to the study, "Health Hazard Evaluation Report," based on workers from the Bloomington Westinghouse plant, see Hinnefeld, 2006. Westinghouse insisted that the correlation between PCBs and cancer did not mean the company was liable. As one spokesperson said in response, "The [NIOSH] study did not link PCBs or any other compound to cancer. It's just a number, a numerical anomaly. It's exposure to sunlight that increases the risk of skin cancer" (quoted in Lane 1991a).

25. Potts eventually was diagnosed with colon cancer, which he suspected was linked to PCB exposure at the plant (see Hinnefeld 2006).

26. According to the American Community Survey Three-Year Estimates of the US Census Bureau (2005–2007), the median travel time to work in Bloomington remains under fifteen minutes, the median household income is $28,540, and 13.3 percent of the population is living below poverty. Home to almost seventy thousand people, Bloomington's population includes 87 percent whites, 8.4 percent blacks, 5.3 percent Asians, 2.5 percent Hispanics/Latinos, and less than 1 percent Native Americans or Native Hawaiians. Bypassing the limestone and electronic industries, Bloomington's largest employer has become Indiana University, with Crane Naval Base following as the second-largest employer (US Census Bureau 2009).

27. "The condition of our own making" is a phrase loosely borrowed from Karl Marx, who Leo Marx (1964, 298) notes is one of the significant voices in history who promoted the idea that "the Age of Machinery transforms men [*sic*] into objects."

References

Agency for Toxic Substances and Disease Registry. 2001. Polychlorinated Biphenyls. Toxic Substances Portal. <http://www.atsdr.cdc.gov/toxfaqs/tf.asp?id =140&tid=26. Accessed: February 7, 2012.

Anderson, Chris. 2005. Flotsam. FlowTV. http://flowtv.org/2005/04/flotsam/. Accessed: February 7, 2012.

Andrews, Mike. 1989. "Radical" Principles Still Govern His Life. Editorial, *Herald Times* (Bloomington), December 28. HeraldTimesOnline.com. Accessed: February 7, 2012.

Beaven, Stephen. 1989. Five Hundred Incinerator Opponents March. *Herald Times* (Bloomington), April 17. HeraldTimesOnline.com. Accessed: February 7, 2012.

Bovee, Tim. 1988. Researcher: PCB Exposure Spans Six Generations. *Herald Times* (Bloomington), April 15. HeraldTimesOnline.com. Accessed: February 7, 2012.

Breaking Away. 1979. DVD. Directed by Peter Yates. Los Angeles: 20th Century Fox.

Bullard, Robert D. 1993. *Confronting Environmental Racism: Voices from the Grassroots*. Boston: South End Press.

Creek, Julie. 1989a. Conditions Unveiled for PCB Incinerator. *Herald Times* (Bloomington), October 6. HeraldTimesOnline.com. Accessed: February 7, 2012.

Creek, Julie. 1989b. PCB Advisory Group Has First Meeting Tonight. *Herald Times* (Bloomington), November 21. HeraldTimesOnline.comwww.heraldtimesonline.com/stories/1989/11/21/archive.19891121.6da29fe.sto. Accessed: February 7, 2012.

Creek, Julie. 1989c. State Tries to Learn If Deaths, PCBs Link. *Herald Times* (Bloomington), November 3. HeraldTimesOnline.com.http://www.heraldtimesonline.com/stories/1989/11/03/archive.19891103.efbcf48.sto. Accessed: February 7, 2012.

Creek, Julie. 1990. Incinerator for Blackwell Stars in Cable Public Service Spot. *Herald Times* (Bloomington), January 23. HeraldTimesOnline.com. Accessed: February 7, 2012.

Douglas, Mary. 1968. *Purity and Danger: An Analysis of the Concepts of Pollution and Taboo*. New York: Frederick A. Praeger.

Fox River Watch. n.d. The History of PCBs: When Were Health Problems Detected? http://www.foxriverwatch.com/a_pcb_pcbs.html.

Francis, Eric. 1994. Conspiracy of Silence: The Story of How Three Corporate Giants—Monsanto, GE, and Westinghouse—Covered Their Toxic Trail. *Sierra Magazine*, September–October. http://www.sierraclub.org///.asp.

Groundbreaking Ceremony Part of Fall Festival's Gala Program despite Rain. 1956. *Daily Herald Telephone* (Bloomington), September 15, p. 1.

Higgins, R. R. 1994. Race, Pollution, and the Mastery of Nature. *Environmental Ethics* 16 (Fall): 251–264.

Higgs, Steven. 1988a. County Gets Help in Its Review of Incinerator Plan. *Herald Times* (Bloomington), June 4. HeraldTimesOnline.com. Accessed: February 7, 2012.

Higgs, Steven. 1988b. List of Suspected PCB Sites Goes to Officials: Banks Nervous about Loans on Affected Property, Site-Search Committee Fears Liability. *Herald Times* (Bloomington), April 1. http://www.heraldtimesonline.com/stories/1988/04/01/archive.19880401.f4003d6.sto. Accessed: February 7, 2012.

Higgs, Steven. 1990. PCB Health Concerns Began 50 Years Ago. *Herald Times* (Bloomington). December 10. HeraldTimeOnline.com.http://www.heraldtimesonline.com/stories/1990/12/10/archive.19901210.204dff2.sto. Accessed: February 7, 2012.

Higgs, Steven. 1992a. Checks for Eight Workers Exposed to PCBs Arrive. *Herald Times* (Bloomington), February 1. HeraldTimesOnline.com. http://www.heraldtimesonline.com/stories/1992/02/01/archive.19920201.361d233.sto. Accessed: February 7, 2012.

Higgs, Steven. 1992b. Payment in PCB Lawsuit Delayed. *Herald Times* (Bloomington), January 5. HeraldTimesOnline.com. Accessed: February 7, 2012.

Higgs, Steven. 1994. Protests Key to Decision. *Herald Times* (Bloomington), February 12. HeraldTimesOnline.com. Accessed: February 7, 2012.

Hinnefeld, Steve. 1988a. EPA to Add 189 Suspected PCB Sites to Its List. *Herald Times* (Bloomington), April 29. HeraldTimesOnline.com. Accessed: February 7, 2012.

Hinnefeld, Steve. 1988b. Incinerator Study Questioned. *Herald Times* (Bloomington), June 21. HeraldTimesOnline.com. Accessed: February 7, 2012.

Hinnefeld, Steve. 1988c. Jury Rules in Favor of Monsanto 7-16-88. *Herald Times* (Bloomington), July 16. HeraldTimesOnline.com. Accessed: February 7, 2012.

Hinnefeld, Steve. 1988d. Monsanto Downplays Risks of Human Exposure. *Herald Times* (Bloomington), July 14. HeraldTimesOnline.com. Accessed: February 7, 2012.

Hinnefeld, Steve. 1988e. Monsanto Says PCBs "Helped." *Herald Times* (Bloomington), July 6. HeraldTimesOnline.com. Accessed: February 7, 2012.

Hinnefeld, Steve. 1988f. 189 Other PCB Sites Won't Go on List Yet, EPA Officials Say. *Herald Times* (Bloomington), May 7. HeraldTimesOnline.com. Accessed: February 7, 2012.

Hinnefeld, Steve. 1988g. Risk Study Termed Useful, Incomplete. *Herald Times* (Bloomington), June 28. HeraldTimesOnline.com. Accessed: February 7, 2012.

Hinnefeld, Steve. 1988h. Thirteen Area Land Owners Consent to PCB Test. *Herald Times* (Bloomington), June 23. HeraldTimesOnline.com. Accessed: February 7, 2012.

Hinnefeld, Steve. 1988i. Westinghouse Knew of PCB Spread in '71: Schalk Gives Report to Board. *Herald Times* (Bloomington), May 10. HeraldTimesOnline. Accessed: February 7, 2012.

Hinnefeld, Steve. 1991. PCB Incinerator Bill Approved by House Committee. *Herald Times* (Bloomington), February 1. HeraldTimesOnline.com. Accessed: February 7, 2012.

Hinnefeld, Steve. 1997. Westinghouse to Begin PCB Cleanup Monday: Sludge, Soil to Be Removed from Winston-Thomas Plant, Clear Creek. *Herald Times* (Bloomington), August 21. HeraldTimesOnline.com. Accessed: February 7, 2012.

Hinnefeld, Steve. 2000. Two Families to Move Due to High PCB Levels During Cleanup. *Herald Times* (Bloomington), July 22. TimesHeraldOnline.com. Accessed: February 7, 2012.

Hinnefeld, Steve. 2006. Westinghouse, PCBs, and Cancer: New Study Sees More Deaths from Certain Types of Cancer among Former Factory Workers Exposed to PCBs. *Herald Times* (Bloomington), February 26. HeraldTimesOnline.com. Accessed: February 7, 2012.

Kuletz, Valerie L. 1998. *The Tainted Desert: Environmental and Social Ruin in the American West*. London: Routledge.

Lane, Laura. 1990a. A Legacy of Worry: Former Workers Face Health Uncertainies [*sic*]. *Herald Times* (Bloomington), December 9. HeraldTimesOnline.com. Accessed: February 7, 2012.

Lane, Laura. 1990b. A Legacy of Worry: On the Line Working with PCBs. *Herald Times* (Bloomington), December 9. HeraldTimesOnline.com. Accessed: February 7, 2012.

Lane, Laura. 1990c. A Legacy of Worry: PCB Workers Pressing Legal Claims. *Herald Times* (Bloomington), December 10. HeraldTimesOnline.com. Accessed: February 7, 2012.

Lane, Laura. 1991a. Health Alert Not Strong Enough, State Says. *Herald Times* (Bloomington), February 20. HeraldTimesOnline.com. Accessed: February 7, 2012.

Lane, Laura. 1991b. Westinghouse Retirees Respond. *Herald Times* (Bloomington), April 2. HeraldTimesOnline.com. Accessed: February 7, 2012.

Litke, James. 1985. EPA Calls Westinghouse Settlement Largest in Hazardous Waste Case. *Associated Press*, May 20.

Livezey, Emilie Travel. 1980. Poison Alert. *Christian Science Monitor*, April 10. Accessed: February 7, 2012.

MacNeil, Robert. 1985. Windfall? Broadcast Battle; Busing in Denver. Host. *MacNeil/Lehrer NewsHour*, May 20.

Marx, Leo. 1964. *The Machine and the Garden: Technology and the Pastoral Ideal in America*. Oxford: Oxford University Press.

Monroe County History Center. n.d. A Short History of Bloomington and Monroe County. City of Bloomington. http://bloomington.in.gov/documents/viewDocument.php?document_id=3052. Accessed: February 7, 2012.

Montague, Peter. 1989. Thanks to Monsanto. *Rachel's Environment and Health News* 144 (August 28). http://www.ejnet.org/rachel/rhwn144.htm. Accessed: February 7, 2012.

Montague, Peter. 1993. How We Got Here—Part 2: Who Will Take Responsibility for PCBs. *Rachel's Environment and Health News* 329 (March 17).

Nolan, Bethany. 2009. Local BCB Cleanup Worries Unfounded, EPA Says. *Herald Times* (Bloomington), October 7. Accessed: February 7, 2012.

PCB Records to Be Turned Over. 1988. *Herald Times* (Bloomington), June 1. TimesHeraldOnline.com. Accessed: February 7, 2012.

Peterson, Cass. 1985. Westinghouse Signs Pact on Toxic-Waste Cleanup; Cost at Six Sites Could Total $100 Million. *Washington Post*, May 21.

Pezzullo, Phaedra. 2001. Performing Critical Interruptions: Rhetorical Invention and Narratives of the Environmental Justice Movement. *Western Journal of Communication* 64 (1): 1–25.

Phillips, Donald L., Virlyn W. Burse, Gregory K. Steele, Larry L. Needham, and W. Harry Hannon. 1989. Half-life of Polychlorinated Biphenyls in Occupationally Exposed Workers. *Archives of Environmental Health* 44, no. 6 (November–December): 351–354.

Reich, Michael R. 1991. *Toxic Politics: Responding to Chemical Disasters*. Ithaca, NY: Cornell University Press.

Risebrough, Robert, and Virginia Brodine. 1971. More Letters in the Wind. In *Our World in Peril: An Environment Review*, ed. Sheldon Novick and Dorothy Cottrell, 243–255. Greenwich, CT: Fawcett.

Schroeder, Michael. 1991. Did Westinghouse Keep Mum on PCBs? *Business Week*, August 12, 68–70.

Shabecoff, Philip. 1985. Westinghouse to Clean Up 6 Toxic Waste Dumps in Indiana. *New York Times*, May 21.

Sheckler, Jackie. 1990. Neighbors Launch Landfill Fight. *Herald Times* (Bloomington), December 11. TimesHeraldOnline.com. Accessed: February 7, 2012.

Sheckler, Jackie. 1991a. The Changing Face of INCINERATOR OPPOSITION. *Herald Times* (Bloomington), April 28. HeraldTimesOnline.com. Accessed: February 7, 2012.

Sheckler, Jackie. 1991b. D'Amico: West Side "Forgotten." *Herald Times* (Bloomington), May 3. HeraldTimesOnline.com. Accessed: February 7, 2012.

Sheckler, Jackie. 1991c. Groups Unite to Fight Incinerator. *Herald Times* (Bloomington), April 23. HeraldTimesOnline.com. Accessed: February 7, 2012.

Sheckler, Jackie. 1992a. ABB Workers Seek to Halt PCB Cleanup. *Herald Times* (Bloomington), June 24. HeraldTimesOnline.com. Accessed: February 7, 2012.

Sheckler, Jackie. 1992b. Workers Still Seek Answers about PCB Exposure. *Herald Times* (Bloomington), February 2. HeraldTimesOnline.com. Accessed: February 7, 2012.

Sheckler, Jackie. 1993. Workers Want PCB Study Expanded. *Herald Times* (Bloomington), August 17. HeraldTimesOnline.com. Accessed: February 7, 2012.

Sheckler, Jackie. 1994a. Incierator [*sic*] Off: PCB Gets Fresh Start. *Herald Times* (Bloomington), February 12. HeraldTimesOnline.com. Accessed: February 7, 2012.

Sheckler, Jackie. 1994b. PCB Decision "Good News for Community." *Herald Times* (Bloomington), February 15. HeraldTimesOnline.com. Accessed: February 7, 2012.

Sheckler, Jackie. 1994c. Workers File PCB Suit: $100 Million Sought in Damages. *Herald Times* (Bloomington, June 24). HeraldTimesOnline.com. Accessed: February 7, 2012.

A Spreading Stain from PCB Pollution. 1976. *Business Week*, May 17, 39.

Steel for W.E. Plant on Hand; Plans Unchanged. 1957. *Daily Herald Telephone* (Bloomington), February 13, 1.

Test for PCB Contamination. 1984. *United Press International*, March 26.

Top Westinghouse Men in Groundbreaking Rite. 1956. *Daily Herald Telephone* (Bloomington), September 12, 1, 10.

Trewhitt, Jeffrey, and Lauri A. Rich. 1985. EPA's Biggest Waste Win Ever. *Chemical Week* (May 29): 12.

Tsing, Anna Lowenhaupt. 1993. *In the Realm of the Diamond Queen*. Princeton, NJ: Princeton University Press.

Tsing, Anna Lowenhaupt. 2005. *Friction: An Ethnography of Global Connection*. Princeton, NJ: Princeton University Press.

US Census Bureau. 2005–2007. *American FactFinder*. http://factfinder.census.gov.

US Census Bureau. 2009. *State and County QuickFacts*. http://quickfacts.census.gov/qfd/states/18/18105.html. February 7, 2012.

US Department of Commerce, Bureau of the Census. 1952. 1950, Vol. 2, Characteristics of the Population, Part 14, Indiana, Table 87. In *Census of Population*. Washington, DC: US Government Printing Office.

US Department of Commerce, Bureau of the Census. 1953. 1950, Vol. 1, General Characteristics, Part 3, Idaho–Massachusetts, Table 17. In *Census of Housing*. Washington, DC: US Government Printing Office.

US Environmental Protection Agency. n.d. Polychlorinated Biphenyls (PCBs): Laws and Regulations. http://www.epa.gov/epawaste/hazard/tsd/pcbs/pubs/laws.htm.Accessed: February 7, 2012 .

Van der Dussen, Kurt. 1991. Bayh OKs Bill Aimed at Slower Incinerator. *Herald Times* (Bloomington), May 16. HeraldTimesOnline.com. Accessed: February 7, 2012.

VH-1 WORLD ALERT. 1990. *Bloomington, IN/Toxics*. Produced by L. Harrington. Edited by D. Leveen. New York: Editel NY.

Westinghouse. 2009. Timeline. http://www.westinghousenuclear.com/Our_Company/history/Timeline/index.shtm.

Winkle, Chuck. n.d. Calculation of the Center of Population for Bloomington, Indiana. City of Bloomington Department of Information and Technology Services. http://bloomington.in.gov/population-center.

6

The Garbage Question on Top of the World

Elizabeth Mazzolini

Under the intense conditions of extreme atmosphere and topography as well as media scrutiny, Mount Everest, the highest mountain in the world, has become an overdetermined icon of the sort that makes it difficult to distinguish between what is "natural" and "cultural" about its identity and status. Of course, Mount Everest is not unique for mixing the natural with the cultural; it is only unique in the particular ways in which nature and culture blend there. Indeed, especially since William Cronon's edited collection of nature-culture studies *Uncommon Ground* was published in 1996, humanistic environmental scholars have found it difficult to imagine how nature might be separate from our culturally laden ideas about it. In the introduction to *Uncommon Ground*, Cronon notes that ideas of nature are always ultimately products of the culture that inscribes them. He identifies some of the embedded concepts of nature that have come to structure various forms of environmentalism—namely, nature as naive reality, moral imperative, Eden, virtual reality, commodity, return of the repressed, and contested terrain. Far from being mutually exclusive, many of these notions would seem to be intimately related and even mutually supportive (Cronon 1996, 23–56).

All of these conceptions of nature can be seen at work on the natural-cultural entity known as Mount Everest. To wit: Everest occupies a place in the cultural imagination as a site of purity and permanence that must be protected from human-produced degradation (naive reality, moral imperative, and Eden). At the same time, the mountain is thoroughly permeated by Western fantasies about extremity and conquering nature as well as making money from access to nature (virtual reality and commodity). And yet Mount Everest, with its often-unpredictable weather along with its shifting ice and rock landscapes, is seen as having a mind of its own, capable of exacting karmic revenge (return of the repressed). Given the multitude of vested interests in the mountain—many of them in conflict with one another, and/or operating under the auspices of defending different ideas of Everest's purity or exploitability—Mount Everest is undoubtedly, even perhaps above all else, contested terrain.

Contestations over Everest become especially poignant when focused on the issue of garbage accumulation—a problem on Mount Everest commonly attributed to capitalism, especially in recent decades, as the highest mountain on earth has become an increasingly popular tourist destination. Without minimizing the seriousness of the problem, this chapter explores the extent to which garbage exists as much in the imagination and minds of people as it does on the ground. Just to be clear, garbage's powerful hold on people's imagination does not make it a "fake" problem, just as garbage's mere presence on Mount Everest is not the full expression of the "real" nature of the problem (and neither does it suggest an obvious solution). An analytic look at examples of nature-culture configurations, such as those in *Uncommon Ground*, reveals that our ideas of nature (and judgments about what should happen to it) actually are founded as much in our values and ideals, including those unconsciously held, as they are in any independent external reality. How those ideals and reality are related depends on the specificities of the situation.

In that vein, this chapter will take a close look at the particular nature-culture configuration known as Mount Everest in order to demonstrate that judgments about the causes and extent of the garbage problem on Mount Everest as well as potential solutions to it owe less to (malleable and open to interpretation) material facts than to (embedded and difficult to dislodge) cultural ideas about what nature is, what it is for, and how it should be treated. These cultural notions create conflicts over how to define garbage, and whether it is something to be blamed on individuals, groups, institutions, or ideological systems—or whether it is even something to be blamed on anyone or anything at all. There are varying levels of certainty regarding facts, assessments, and policies surrounding the garbage problem on Mount Everest, and that certainty (and uncertainty) reveals more about financial, affective, and collective archetypal investments in the mountain than it does about external reality. These investments tend to focus on entitled individual consumerism, and eschew systemic or nonsubjective approaches to Everest's environment. This is not to make the now well-rehearsed point about the cultural construction of facts but rather to show how, in this one instance, that cultural construction actually constitutes reality, and conversely (after Cronon) that reality can only be known through cultural constructs—the dominant of which is, on Mount Everest, capitalism and the contradictions it exploits to make people feel good about themselves.

It is only in recent years that people have come to strongly associate Mount Everest with garbage, and those associations are usually about litter—garbage cast off by individual careless people. Although it was in a 1963 issue of *National Geographic* that the phrase "the world's highest junkyard" first appeared (as a caption for a photo of litter), it wasn't until the 1990s that the concept took hold of the cultural imagination, and the phrase and its variations (like "the world's highest garbage dump") enjoyed widespread use.[1] Cynicism about what a growing climbing industry would do to the presumed purity of the world's highest mountain—previously perceived as formidable, but increasingly viewed as accessible to anyone who could afford the price of admission—led many commentators to assume that the place was strewn with the litter of entitled paying clients of exploitative guide firms. Such claims are difficult to ascertain. The mountain does see a lot of litter during climbing season, but because many guide firms' businesses depend on bringing people a pleasant Mount Everest experience, and because there are other government and industry incentives to keep the mountain clean, there are good reasons to believe that eyewitness accounts only captured one fleeting moment and did not represent the mountain's overall constant state (Bishop and Naumann 1996). There have also been so-called green expeditions, however, whose stated goal has been to clean up the mountain, and that have fallen short of that aspiration or betrayed their stated principles.

In any case, right now, because of individual accounts and cynicism about capitalism's ravaging nature, the idea that Mount Everest is strewn with garbage appeals to common sense. Yet given that this judgment-laden commonplace is beholden to unexamined assumptions about the status of nature as well as human behavior with respect to nature, it deserves further scrutiny in order to better understand how Everest's garbage problem has been produced within a network of cultural investments as deeply embedded and nearly invisible as are ideas about nature. With the garbage question on Everest focused mostly on litter, these cultural investments seem to be structured by entitlement, individuality, responsibility, and consumerism—all imports to Everest in the same way that conceptions about its purity, degradation, extremity, and exploitability also are imports.

Mount Everest's Cultural Constructions

In a sense, there could never have been a Mount Everest (of the sort we have come to know) without Europeans along with other Westernized and industrialized people wanting, trying, succeeding, and failing to climb it. This geographic extremity was explored (relatively) shortly after it was measured and named, and quickly conscripted into human values, goals, and economies. Ascertained in 1852 to be the highest mountain on earth by the Great Trigonometric Survey, a mapping endeavor of the Indian subcontinent meant to serve British colonizing interests, Mount Everest was named for the survey's leader, and took immediate hold of British and European mountaineers' imaginations. After reconnaissance missions in the late nineteenth and early twentieth centuries, concerted efforts to reach the summit began in the 1920s. While many European nations had their sights set on conquering the earth's heights, for various political and geographic reasons Mount Everest became known as the "British" mountain. After failed attempts by British military people and patriots in the 1920s and 1930s, including some well-known tragic endings (such as the storied fate of George Mallory and Sandy Irvine, who disappeared on Everest in 1924—a tale familiar to many armchair mountaineers), finally in 1953 the summit of Mount Everest was reached. Although the overall expedition that supported this first achievement of the summit was British, the two people who ultimately made it were not British nationals. Edmund Hillary was a New Zealander and Tenzing Norgay Sherpa was a citizen of India (although ethnically Sherpa). Their feat, celebrated along with the coronation of Elizabeth II of England, nevertheless marked the close of an era of patriotic mountaineering. This is to say that after 1953, people began to climb less for the honor of their country, as British mountaineers in the 1920s and 1930s did, and more for themselves.[2]

This trend intensified when, in the late 1980s, guided tours to Mount Everest started. Whereas previously the people who climbed Mount Everest had been national heroes, midcentury coverage saw them as rugged individuals, and then in the late twentieth century the people who climbed Mount Everest became consumers—people whose mountaineering skills mattered less than their ability to pay for a trip to the top. Over this period of time the ethos of the mountaineer became increasingly distilled to something resembling entitled egoism. Thus the twentieth century witnessed Mount Everest's transformation from a nationalist trophy, to a reward for individual wherewithal, to a consumer good to be bought and sold on the open market (Kodas 2008; Rogers 2007).[3]

Thus, Mount Everest as we now know it has been, so to speak, brought into existence by Western values, institutions, and processes. Mount Everest's cultural construction is one fundamental ambivalence in Everest's Westernized existence, because Mount Everest would seem to be the very antithesis of culturally constructed. At 5.5 miles above sea level, rising through several layers of atmosphere, Mount Everest's geologic proportions are immense and cannot be overstated. Additionally, with shifting icefall landscapes and weather that allows for human presence only two months of the year (its weather is unpredictable even during those scarce "safe" weeks anyway), Mount Everest's conditions are nonnegotiable, and have been surmounted only after great pains, many casualties, and a lot of luck. The mountain's status and value in culture over the past century or so, notwithstanding its overwhelming indifference and immensity, has shifted according to human events. In spite of all the ways it exceeds human time, scale, and effort, it has nevertheless come to seem that humans can bring Mount Everest under control and manage it according to their interests. Even as the mountain has had this humanly accessible profile thrust on it, people have idealized Everest as a formerly pristine natural object increasingly debased by human presence.

Mount Everest's apparent contradictions do not end there. While it seems as though the present and immediate environmental threat to Everest is the present and immediate litter on the mountain left by paying visitors, unseen forces could actually be even more threatening. In addition to the mountain's conflicted identity as both inherently natural and thoroughly cultural, its huge (and still growing) popularity as a tourist attraction seems like it might pose a problem, because it appears obvious that a lot of traffic would lead to environmental degradation. Looked at from a different perspective, however, Mount Everest's and the Himalayas' contemporary problems might be seen as linked less to tourists and climbers, even in large numbers, than they are to much wider-scale (that is, global) industry, which is undermining Everest's environment rather than merely sullying its landscape. Climate change is causing Mount Everest's ice to melt.[4]

In fact, even the garbage problem right now might not be because of the careless habits of current climbers but rather because climate change is unearthing the past. One leader of a recent environmental expedition notes that past decades' garbage, once buried under snow, is emerging as the snow melts because of global warming.[5] The past accumulation of garbage, whose main form of degradation seems to be unsightliness, has only become a problem with the onset of the much bigger and more damaging issue of global climate change. According to reports, the Khumbu icefall, a famously dangerous maze of shifting ice cliffs that is regularly negotiated by climbers on one of the more heavily trafficked routes up Mount Everest, has been breaking up as the ice melts due to climate change (Buncombe 2009). In other words, from this albeit-counterintuitive perspective, the tourists visiting the Himalayas are not the mountain's main problem; climate change brought on by distant industries is. This, along with Everest's simultaneously thoroughly natural and cultural identities, is another counterintuitive idea that is difficult to resolve with what feels like a commonsensical notion that people paying to visit Everest are bad for the mountain.

Critics (especially ones using some variation of "the world's highest junk yard") who blame Himalayan degradation on tourism may miss the global point because they are too focused on what is right in front of them. Environmental issues that most profoundly affect the Himalayas take place on a much larger and less localized scale, and with much more devastating effects than can be accounted for by careless and messy tourists, no matter how numerous. And yet for reasons overlooked by many people who are focused on litter, Mount Everest may indeed be under threat because of human activity—just not human activity that can be identified and condemned on-site. Garbage is a question on Mount Everest both because it might *not* be there (given the economic incentives to remove it and untrustworthiness of momentary individual eyewitness accounts), and because even if it is there, it might not be the kind of problem people represent it to be. Moreover, from the global climate change point of view, concentrating on garbage distracts from the more devastating concern. Everest's irreconcilable natural and cultural inconsistencies allow for this distraction, due to the way in which Everest's profile as a pure and natural entity denigrated by human activity on it discourages a view of Everest as part of a global ecology that also includes industrial activity. Our embedded ideals of nature have imposed a locus of responsibility for Everest: the individual human.

The idea that the nature of Everest is a cultural production within a global ecology is a hard pill to swallow for well-intentioned people who claim to want to accept the mantle of responsibility, like those participating in and funding "Eco Expeditions" to Everest. Underwriting these expeditions is the notion that participants are in control of keeping the Himalayas sublime and intact. The same large-scale industrial practices, though, are melting snow and ice all over the world, including the Himalayas. These industrial practices are not located near the places they are affecting most devastatingly, so that trips to these places to pick up litter seem unlikely to really address the issue (Connor 2010).

Although they constitute different groups with different interests, critics who assign Himalayan degradation to contemporary tourism have something in common with so-called climate change skeptics who cite cool weather outside their own houses as evidence that global warming is not occurring. Both groups cite the most immediate, obvious evidence to an individual observer rather than more closely examining systemic issues. For people concerned with the Himalayas, judgment about the situation comes first, in the form of the embedded, natural-seeming idea that nature is pure and becomes sullied by capitalism, followed by the most easily, subjectively acquired evidence that fits the judgment well. Unlikely associates though they may be, the global warming skeptic and critics claiming Himalayan degradation is caused by tourism are part of the same problem: they are stuck on the local and subjective, instead of engaging with the global and systemic.

Those who bring attention to what has come to be called "backpack fatigue"—the wear and tear that popular tourist sites show as a result of hordes of people visiting each year—and ecological tourism, during which the visitors famously are supposed to take only pictures and leave only footprints, are doing important work. Yet it seems possible and even probable that on Mount Everest, these tourists, even in aggregate, are not the biggest issue the mountain faces. Everest is a culturally configured natural object, and therefore the product of human judgment and values about what nature is as well as how it should be treated. Those judgments and values shape the perception of evidence to fit often-foregone conclusions. In this case, the judgment from the beginning is that Mount Everest is pristine and needs to be protected from the hordes of tourists that crass commercialism has let in. The suggestion that careless tourists dropping energy bar wrappers are not a significant problem on Everest seems nearly unthinkable in light of ingrained ideas about nature needing to be protected from individualist culture. In other words, there would appear to be (almost) no popular precedent in thought or discourse, no common communicative or policymaking formation, allowing for the assertion that environmental problems in the Himalaya and especially on Mount Everest might result from anything other than (bad) people making bad choices on-site.

When it comes to tourism and the environment, activist tourists and tour leaders alike are likely to assume the situation is subjective, by which I mean subject based: to be understood in terms of subjective perception and individual responsibility, and to be handled with informed choice and planning. The challenge is to come up with an environmentalist epistemology that is reflective about its values and yet not subject centered.

Narcissism and Venality

Environmentalists in general appear to have that challenge, but it seems like an especially tricky problem on Mount Everest, where thanks to the intensely particular symbolic and material conditions, subject centeredness is all but inevitable. The garbage problem on Mount Everest is rooted in how people have been enabled to act. The cultural situation that allows for various claims to be made (and blame to be laid) about garbage and Everest's environment is structured around the centrality of the individual subject—a situation that is strongly encouraged through common practice even while it is explicitly condemned. Climbers are often indicted for indulging in self-obsession and what is all too easily called narcissism.

One social science study links the well-publicized Mount Everest disaster in 1996 (during which eight people died, chronicled in Jon Krakauer's best seller *Into Thin Air*) to the specific, recent culture that has formed among climbing leaders and clients (Elmes and Barry 1999). This culture, according to the study's authors, is the direct result of the commodification of high-altitude mountaineering, making for climbers who are less skilled and more dependent, and leaders whose primary role is to seek publicity and indulge client desires in order to earn a living. Unlike previous eras, such as during the early and mid-twentieth century, when expedition leaders' jobs were to ensure safety and order while climbing under group auspices, climbing guides are now service providers paid to "deliver the goods"—the summit of Mount Everest—to individual customers. Paying client-climbers feel entitled to this reward, in the way that consumers feel entitled to get what they pay for. The study goes on to describe this consumerist culture as promulgating narcissistic behavior during expeditions, and characterizes the actions of people during the 1996 disaster as consistent with clinical narcissism.

Although no teams since 1996 have faced the exact circumstances that lead to the deaths during that year, the culture remains in place, characterized by structured profitability, competition, and self-interest. As with littering, however, this is not simply a matter of people being drawn to Mount Everest and then choosing ex nihilo to behave badly. Instead, on Everest a complicated system of mutually reinforcing conditions, values, and practices produces individual subjects with a limited range of possible as well as desirable decisions to make. This is yet another complication for Mount Everest: systems thoroughly permeate its status and meaning, and yet individuals emerge as sole moral agents, even though the individual behavior has been made possible and legible only by economic and social systems.

A focus on the individual at the cost of attending to the system serves the interests of profit. Capitalism creates and thrives on the ideology of the special, intentional, and entitled subject that shapes so much human activity on the mountain. In the process, capitalism disappears, and seems natural and transparent, not something to be questioned. Consumerist individuality, too, becomes naturalized. One way to begin to formulate a non-subject-based environmentalism could be to think of this ideology of individuality as an *effect* of capitalism as opposed to a foundational category. This is to say that in the search for systemic instead of subjective reasons for environmental degradation, environmentalists might consider that capitalism produces entitled individuality, rather than that capitalism and its attendant individuality is simply the state of things, and people thus are required to make morally upstanding decisions (even though such decisions are discouraged by a capitalist consumerist system).

Further, as a twist on Cronon's embedded conceptions of nature by applying it to similarly embedded naturalistic conceptions of culture, it would seem worth questioning the embedded conception that capitalism is a given and immutable (that is, natural) part of culture. The capitalism on Mount Everest, linked as it is to entitled consuming individualism, helps maintain a myth of individual praise and blame. When it comes to garbage and environmental degradation, this myth keeps the focus on litterbugs and liars, rather than industrial systems and global effects. Capitalism's individual entitlement and responsibility can seem just as real and immutable as the natural world appears to be. And yet if, thanks to scholars like Cronon, we know the natural world to be the creation of human values, surely the cultural world, emergent as it is from ideologies and artifacts, must be such a production as well.

In his book *High Crimes: The Fate of Everest in an Age of Greed*, journalist Michael Kodas takes the individual moral indictment route (as the title might indicate), but in doing so also perhaps inadvertently points to some of the systemic conditions that have contributed to Mount Everest's current status as a site of irresponsible behavior. He documents many of the ways in which Mount Everest and its immediate environs are lawless lands where people thrive by lying, cheating, and stealing. Kodas (2008, 276) quotes fellow journalist Finn Olaf's dispatch to Discovery.com in the year 2000:

I have stayed at many base camps throughout the world, but Everest's is much different. The money, for starters. There's way too much of it floating around like a cloud on the mountain, and that attracts some people who are more focused on cash than climbing. Then there's the mania that comes with breathing thin air for too long. And the oversized egos and bruised personalities seeking some kind of completion amidst this soaring scenery.

The mingling of nature and culture that produces a certain version of individuality is manifest here, as is the intense motivation for all the money to be made by entrepreneurs. The profit motive goes hand in hand with identity enhancement on Mount Everest. And there seems to be no end point for either—only an imperative for more money and more ego satisfaction. Money on Mount Everest is made running expeditions and attracting those expeditions' clients, who pay upward of seventy thousand dollars per person, depending on the guiding package and included amenities. Corporate sponsorship of expeditions attracts clients, generating more profit in the form of advertising dollars and equipment supplies, and can help make expeditions feel luxurious and high profile for clients already paying exorbitant prices, and who therefore feel entitled to good food, comfortable accommodations, and communication access, not to mention the summit of Mount Everest.

With profit as the primary impetus, some unscrupulous expedition leaders lie about having corporate sponsorship in order to attract paying clients. Kodas alleges that well-known expedition leader Robert Hoffman did just that on his 2000 Everest Environmental Expedition by falsely claiming Discovery Channel sponsorship. (Olaf's discovery of this lie partly motivated his cynical comment quoted above.) Hoffman maintained that one purpose of his expedition was to "clean up" Mount Everest, but it became clear that environmentalism was far down his priority list—well below garnering attention, clients, and profit—and that for Hoffman, environmentalism was an empty signifier to be used as bait for people eager to feel good about themselves while fulfilling a personal goal.

Although Kodas documents all manner of venal and violent individual behavior on Mount Everest and at base camp, his portrayal of untruths about environmentalist intention there suggests systematic propagation supported by the profit motive more so than it indicates that bad people do bad things. On Everest (and perhaps elsewhere) it would seem that profit is a more effective motivator than truth, and when relations between people are structured by profit, as those between expedition leaders and clients on the mountain are, it simply does not matter what the truth about litter or the environment is. As Hoffman knew, offering people ways to construe themselves as accomplishing, righteous, and authentic is more profitable than acceding to dismal and boring truths, such as that climate change and global capitalism are more damaging to Mount Everest than are litterbugs. A non-subject-based environmentalism is up against capitalism, which on Mount Everest thrives because of the lures of subjectivity tied to entitled accumulation. The business models and profit motive distract from themselves as the root of the problem by installing the consuming individual who is then poised to take the blame for ruining Mount Everest. And the blamers are just as much a part of the subjective, capitalistic environmentalism as are the blamed.

Hoffman did not decide that what gets rewarded on Mount Everest is selfhood. Unethical though his actions may have been, Hoffman was working within the system as he found it. As people seek credit and experience for themselves on Mount Everest, cultural and natural conditions continually inscribe them as not only discrete, planning and choosing subjects but also as practically sui generis selves, capable of accruing cultural and material resources against all odds. Climbers on Mount Everest are usually seeking the edification and self-definition that comes with being one of the few people who have stood on top of the world. On this mountain, people pursue earthly and existential extremity simultaneously. The summit of Mount Everest's history in the last half of the twentieth century has been a proving ground for a litany of identity-conferring firsts, such as the first woman, the first person from a certain country, or the first blind person, to cite a few examples, not to mention the ever-receding horizons of being the youngest or oldest person to climb.[6] The importance to climbers of a singular, exceptional existence is also apparent from the tone of the numerous climbers' published memoirs, which reflect grandly on the nature of living (for bourgeois Westerners) and often make cliché-sounding self-help points that sound disingenuously humble.[7] Narcissism sought and identified by researchers is par for the course on Mount Everest; it is also one of the main challenges to systemic environmentalism that posits something other than a willful individual as independent moral agent.

Selfhood and Disgust

If Everest's connection to global climate change exists in tension with much of the culture that maintains Everest as a place for special and blameworthy individuals, yet another environmental threat exists in tension with the purity and rarefication that individuals supposedly achieve on reaching its summit: human waste. Human waste's nearly unspeakable nature further demonstrates how the issue of garbage on Mount Everest has been rendered exclusively as litter and therefore a subjective issue, about blaming people who produce it and praising people who pick it up. In a different way from climate change, human waste shows that environmental issues on Everest are systemic—deeply local and inevitable, beyond the realm of judgment. As Gay Hawkins (2006, 45–70) observes, the moment shit stops being a personal and private issue, it becomes a public one, no longer concerning how people feel about themselves as individuals, but rather seen as a widespread health threat.

Here is an example of how human waste enters into environmental discourse about Mount Everest in a startlingly dire manner that nevertheless gets overlooked. In May 2009, the man who has set foot on the top of Mount Everest more times than any other human being did so yet again, for the nineteenth time. Apa Sherpa has been working as a professional mountain porter and guide since 1985, and has reached the summit every single year since 1990. In 2009, Apa was climbing as part of an Eco Expedition to Mount Everest, staged to draw attention to the effects of global warming on the Himalayas. Before climbing, Apa commented to the press,

The beauty of Everest is deteriorating as climbers leave their garbage on the mountain. . . . We must discourage such practices. The mountain is not just a god for us but the snow and ice is the source for water we drink. Leaving behind human [bodily] waste is not just insulting to the mountain god but also contaminates the source of water. (quoted in Buncombe 2009).

In light of Kodas's points about how environmental expeditions seem more about publicity and money than about the environment, Apa's remark about Everest deteriorating can be viewed cynically as a shameless plug for his expedition. Yet his blunt observation about human bodily waste is a bit surprising even amid cynicism. Including bodily waste in a discussion of waste on Everest interrupts the righteousness and choice associated with environmental cleanup expeditions, and so it rarely gets mentioned in contemporary cleanup efforts. Bodily waste cannot be linked to willful selfhood. Its sheer material democracy and disgustingness along with its manifestation of necessity, negligence, and repression make its presence in internationally popular natural settings especially, insistently unpleasant—a reminder, just as edification is nearly attained, that life is relentlessly mundane. More so than the litter brought down by ecologically minded members of environmentally themed expeditions to Mount Everest, however, human waste has the potential to be an environmental and health hazard, as the quote from Apa reminds us. Soda cans, energy bar wrappers, and oxygen bottles may look ugly, but it is feces and urine that could ruin the drinking water that inevitably flows downhill.

The 2009 expedition that Apa accompanied reportedly brought down five tons of garbage. Understandably, the expedition was reticent in its official reports about the amounts of human bodily waste included with the trash and rubbish that the climbers retrieved, preferring to highlight big-ticket garbage items with less of a gross-out factor. The expedition leader, Dawa Steven Sherpa, reported that the expedition carted down over five tons of mountain trash, including parts of a crashed helicopter, old ropes and tents, ladders, metal cans, and climbing gear (Agence France-Presse 2009). The reticence to talk about one of the most fundamentally defiling substances is no doubt due to the fact that removing human waste is rather disgusting to consider, even when the act essentially resembles merely shoveling ice (Willis 2010). Given the danger to drinking water that bodily waste poses, the widespread reluctance to discuss it is just as dangerous as using environmentalism as an empty slogan to sell things is damaging. In other words, the repressed subject is as harmful as the righteous or profiteering subject.

There is another reason why the Eco Expedition and other environmental expeditions concentrate on garbage in the form of objects cast aside, which might be unseemly but are not necessarily dangerous, instead of human waste, which is nearly invisible yet potentially quite dangerous. Objects tossed aside fit with the conception of Mount Everest as a place where people accomplish things because of their strength and wherewithal, or fail to do so because of their weakness or negligence. Objects cast aside, in short, align with the conception of Mount Everest bearing on climbers' personal qualities—as narcissists or do-gooders, for example. Bodily waste does not cognitively fit with the environmental situation on Mount Everest, because bodily waste is not clearly linked with other environmentally significant behaviors—like littering or overconsumption—that can be observed, managed, and judged. It cannot be connected to strength, wherewithal, weakness, or carelessness, and while it can in many cases be managed, it can never be altogether suppressed.

The special entitled consuming individual who climbs Mount Everest is partly a product of a repressive discourse about this gross, uncontrollable substance. Aside from the consumer culture that encourages people to think of themselves as elite and intensely special individuals when they climb this mountain, there are some aspects of life at high altitudes that help obfuscate the mundane facts of life—a case of natural and cultural conditions reinforcing each other. With its rarefied and dry atmosphere,

Mount Everest fosters none of the visceral effects of uncanny otherness from which people often seek to distance themselves when they are confronted with as well as disgusted by the facts of life. Physical conditions on Mount Everest simply are not conducive to producing disgusting substances or objects. The atmosphere cannot harbor moisture, fecundity, or decomposition, so that none of the textures, sights, or odors usually associated with the disgusting can exist there.[8] (Threats to the drinking water flowing downhill come in a frozen, dry, powdered form.) Also, thanks to systematic and widely used toilet systems, and the broad economic incentives for keeping the mountain clean, the substances that Apa talked about as defiling the mountain do not feel present to those ambitious and paying visitors seeking to enhance their identity, and who therefore must deal with the quotidian aspects of human bodies on Mount Everest barely any more than they do at home. The natural, material world on Everest fits right in with cultural, ideological values.

Visitors can concentrate on producing themselves as rarefied and elite, to match their fantasies of the mountain. This process is made possible by repressing defilement and danger while publicizing good deeds and impressive accomplishments, as the Eco Expedition did. Because the Eco Expedition ultimately did not report whether and how it addressed any of the human bodily waste on the mountain, the material (water contamination) and spiritual (insulting to the mountain god) effects of it as voiced by Apa were left unaddressed publicly. Paying clients, who would most likely drink bottled water and not practice the Sherpa religion, would not experience these effects. Of course, Apa's comment about bodily waste might still be viewed cynically, as a dramatic illustration of the environmental dangers he was getting paid to help ameliorate, but it also might be seen as an intrusion of something beyond human control and judgment, and hence startlingly out of keeping with the status quo on Mount Everest.

The fact that human bodily waste represents a threat that is disgusting and dangerous, yet ambiguous and seemingly unreportable, is one more way that garbage and waste are in question on Mount Everest. Moreover, the cultural and natural mechanisms in place that keep human bodily waste simultaneously disgusting, dangerous, and ambiguous suggest further that the question of garbage results from the way systems of knowledge production surrounding Mount Everest work. These systems make legible only those substances as well as activities that can be attributed to individuality, entitlement, and righteousness, in contrast to those that are systemic, public, or beyond the judgment of individual behavior.

Ecoironies and the Consequences of (and Possibilities for) Uncertainty

Everest's cultural contradictions may be lodged in the incommensurability between the systemic, public, and individual, but they do not end there. Environmentally minded and themed expeditions like Eco Expedition 2009, which Apa accompanied, bear markers of hierarchies and status, because people (and corporations) like to have their names and identities associated with high-profile good behavior. Even though environmental expeditions are ostensibly there for cleanup, members usually want to also try to reach the mountain's summit, even though there is less garbage the closer you get to the top since there are fewer people who make it that far. It is much more impressive to have reached the top of Mount Everest (and lived to tell about it) than it is to have picked up some litter there. It is even better if you can do both, and get credit for being both physically hearty and morally righteous. Such was the case with the Eco Expedition in 2009 that Apa accompanied—many members of which did reach the summit.

Concerted efforts to reach the summit of Mount Everest require a lot of time and resources beyond even those required for a stay at base camp, and so can come at the expense of other purported goals on the mountain. Within the systems of value on Mount Everest, it would be a relatively easy calculation if one had to decide between allocating resources for either picking up garbage or reaching the top. Anyone can pick up a Coke can, but the overwhelming cultural message is that it takes a special someone to haul oneself into the stratosphere. It is worth leaving the Coke can behind while leaving time and energy to try to reach the top. Such are the cultural contradictions surrounding Everest's status: pure, but there for human use.

As Kodas suggests, there are many expeditions climbing under the auspices of environmentalism; the Eco Expedition hardly has a monopoly on cleanup. There are some ironies involved with such expeditions. If, for instance, as some environmentalists and mountaineering purists point out, the environmental problem on Mount Everest is defined by the growing number of tourists, with their inevitable garbage and waste production, it seems odd to bring yet more people on yet another massive expedition. It is likewise strange to further contribute to the glamorous and elite mystique that surrounds Everest with a highly publicized, record-breaking, virtuous expedition, thereby making the mountain and its summit seem even more appealing to more people who will presumably litter.

The questionable status of garbage on Mount Everest has additional consequences of interest to those who study the relationships between the material and symbolic aspects of environmentalism. Not only is there a history of the garbage question on Everest, partly represented here, but the questioning itself has produced some specific consequences too. One such effect is the trashing of uncertainty because it is not profitable. The material existence and symbolic significance of garbage on Mount Everest might be uncertain, given that it seems to be less damaging than it appears at first glance, and given how squarely it plays into the dominant but culturally laden sentiment about Everest's status and role. Yet it is certainty, even if feigned or disingenuous, that gets publicity and profits. Thus, the certainty about both pounds of litter removed by Eco Expeditions like Apa's and Hoffman's lies relating to his own expeditions is crucial for maintaining the ideology that people know what is going on and what should be done up on Mount Everest.

Just as including either a systemic focus or concerns about bodily waste in environmental discourse about Everest could interrupt discourses of entitled subjectivity, so too could embracing uncertainty interrupt epistemologies of righteous individualism. The question of garbage on Mount Everest has been something profiteers like Hoffman can capitalize on, but uncertainty could also contribute to a non-subject-based environmentalism if it were to be embraced rather than exploited. Uncertainty has valuable and productive possibilities, and yet an explicit state of uncertainty seems to be precluded by ideological conditions in much the same way as disgust is precluded by material and cultural conditions on Mount Everest. New nonsubjective knowledge formations could assess and address the environmental impacts of human activity in the high mountains, and embracing uncertainty could help create such new knowledge formations.

Certainty might not be all it's cracked up to be anyway. Even the scientific method, with its vaunted neutrality and principles that allow for abstraction and applicability, might not be suited for the task of creating certainty, since science inevitably has vested interests and reveals some facts at the expense of obscuring others. *Uncertainty on a Himalayan Scale* by Michael Thompson, Michael Warburton, and Tom Hatley (1986, 6) evaluates environmental issues in the Himalaya, and observes that

the uncertainties surrounding the key variables in the man-land interactions in the Himalaya (and, worse still, the uncertainties as to what the key variables *are*) render the "the problem" unamenable to the traditional problem-solving methods of applied science. The problem, we conclude, is that there is not *a* problem but a multiplicity of contending and contradictory problem definitions, each of which takes its shape from the particular social and cultural context that it helps to sustain. The problem, in other words, is *trans-scientific* in nature.

Traditional, scientific problem-solving methods are ones that assume adequate foreknowledge, abstractable principles, and carefully managed planning—activities of discrete and self-determining subjects.[9] When these authors describe the problem as transscientific, they are calling for altogether new conceptions of who produces knowledge and how. The same instrumental nature-conquering approach that led humans to the heights of the Himalayas in the first place probably will not work to fix any problems.

Thompson, Warburton, and Hatley (1986, 74) go on to advocate for a "multiple problem/multiple solution" approach, in which numerous conflicting, even contradictory definitions and actions are embraced as well as pursued, with, as they put it, "heterogeneity at every level." The authors point to issues in the Himalayas at numerous local, informal, political, institutional, national, and international levels about which there is significant disagreement. Probably one of the most vivid and illustrative of these disagreements is the fact that estimates of per capita fuel wood consumption in Nepal differ by a *factor* (not a percentage) of sixty-seven. Thompson, Warburton, and Hatley want to engage the interests that lead to such a disagreement (rather than discredit those who led to a conclusion with which they disagree), and accept uncertainty, disagreement, and heterogeneity as inevitable and useful. I would add that uncertainty, disagreement, and heterogeneity are useful because they interrupt a unified individualized subjectivity—the basis for certainty and righteousness.

It would seem that when it comes to Mount Everest and the Himalayas, one could exploit or embrace uncertainty. Jack D. Ives, adviser to the United Nations on sustainable mountain development, is another activist and author who advocates embracing uncertainty—this time in order to undo the harm that decades' worth of certainty has caused in the Himalayas. In *Himalayan Perceptions*, Ives (2004) traces the disputes among experts about the state of the Himalayan environment to the theory of Himalayan environmental degradation, which was constructed and propagated (with great moral and epistemic certainty) over the course of the 1970s and 1980s, and into the 1990s. According to Ives, this theory was based on incorrect interpretations of information and images, and epistemology that most valued the knowledge of experts with investments in nongovernmental organizations and scientific institutes as opposed to the local areas they studied.

In *The Himalayan Dilemma*, Ives and his coauthor Bruno Messerli (1989) point out that such expertise serves the interests of only the sponsoring organizations (by seeking only those facts that fit their mostly foregone conclusions), while ignoring or otherwise discounting the knowledge and experience of, for example, indigenous farmers. Ives underscores the ways that scientists and nongovernmental organizations even actually blame farmers for environmental degradation, by claiming that they are responsible for deforestation (recall the disparate estimates of fuelwood consumption), in spite of Ives's evidence that the opposite is true—that farmers are in fact responsible for much of the forest retention and replanting that has gone on over the past several decades (ibid.).

The picture that Ives and Messerli paint of the indigenous farmers in the Himalayas is different from the one many people involved with mountaineering on Mount Everest would depict. The accepted narrative, which suits the interests of those with money and power invested, and aims for moral certainty, is that visitors from developed countries mountaineering in the high Himalayas has been good for indigenous people. According to many experts not invested in the Everest guiding industry, it is still unclear whether tourism is a good or bad thing for Mount Everest and the Himalayas (Rogers 2007). What is all too obvious, however, is that ideologies of individuality, entitlement, and certainty are rampant in the region, such that the large-scale industrial and political effects on the environment are ignored, while the profitability of affective attachments to praise and blame remains powerful. Contemporary environmental discourse about Mount Everest suggests the firm hold ideology has on epistemology. This hold might be broken not by simply countering facts but by interrupting its material logic instead.

Green Gone Wrong: How the Economy Is Undermining the Environmental Revolution, by Heather Rogers (2010), does not mention Mount Everest or the Himalayas, but in its concluding chapter notes that the misinformation, backfiring solutions, and corruption that Rogers details regarding so-called green segments of certain industries are the products not of corrupt individuals but rather of the very nature of capitalism. Rogers does not seek to indict particular pockets of bad actors; she wants instead to show that those bad actors have been produced within the system they have to work with, and that it is the system that needs to be corrected versus blaming or punishing those actors whose choices have been produced as well as constrained within the system. I wish to make a similar point about the Mount Everest climbing industry: its entanglement in profit based on people's identities and feelings about themselves is the problem. Moving away from moralistic judgments means moving away from the tight hold capitalism has on the region.

Interrupting the System

Uncertainty in the Himalayan context seems to be an inevitable and undesirable product of human presence, related to systems rather than individuals, without a discrete locus of blame, and unable to be behaviorally managed with injunctions, much like the human bodily waste—assiduously avoided too—that I mentioned earlier in this chapter. In accordance with dominant interests, uncertainty has come to be not just trash but also shit knowledge—undesirable and repressed, even though universal and inevitable. Subject-based environmentalism continues this repression at our peril, and affirming the inevitable and paying more attention to what does not fit into neat structures of choice and behavior could gain more purchase on the goal of aligning the diverse interests of the earth with those of its inhabitants. Uncertainty, in other words, could be a mode that interrupts the entitlement-based subjectivity that prevails on Everest.

The stories of trash on Mount Everest are also stories of trash knowledge; because of conflicting interests and information, there are a lot of claims that seem useless, but that are repurposed to yield profit. For example, as I have shown, the assertion that litter is a problem on Mount Everest is in question, but nevertheless it is used to attract paying clients and corporate sponsorship. The story of trash and trash knowledge in the Himalayan region is not a tale of the mere suppression of knowledge in the interest of capitalism; it is about extracting cash value from every available niche, and transforming even neglected or murky bits of knowledge into moneymaking machines. With financial success dependent on how people feel about themselves when they visit the mountain, people and corporations with money to make have little to lose and a lot to gain by affirming whatever paying clients want to believe because it makes them feel good about themselves. On Mount Everest, all knowledge is trash knowledge, which is also to say that all knowledge is potentially profitable.

Pointing out the falsehoods and contradictions of environmentalist claims only continues the logic of subject-based praise and blame that at least distracts from the "real" problems on Everest and in the Himalayas, and at most worsens them. A more useful approach might be to modulate some of the practices that attach environmentalist claims to the profitability of entitlement and certainty—by eliminating, for example, Eco Expeditions and other efforts that focus exclusively on litter. On Mount Everest, entitlement and certainty are profitable, which should make them automatically suspect, yet the logic has come to seem commonsensical. It could nonetheless be redirected by strategic counterpractices that would disengage environmentalism from subjectivity. This would involve practical systemic interventions instead of ideological conflicts and disputes over exactly how much litter sullies the mountain, whose fault that litter is, or how misleading expeditions' promises are.

Currently, environmentalist claims on Mount Everest—whether true or false—only gain purchase because of the mountain's imbrications with the subjectivities of climbers and would-be climbers, and because those subjectivities are exploitable for profit. Many of the environmental facts regarding Mount Everest and the Himalayas are also debated, obfuscated, obscured, or otherwise uncertain, which leaves plenty of room for capitalistic interests to take hold and make money, because truth is secondary to personal accomplishment and profitability. My chapter has not been meant to arbitrate the debates about the presence or significance of garbage on Mount Everest but rather to show how seemingly opposing sides are mired in similar ideologies, and suggest that while agreement, certainty, and good conscience seem to be worthy goals in the service of environmentalism, these modes can actually often reproduce the same ideologies that cause problems to begin with.

Notes

1. The original usage is in Bishop 1963, 489. For variations on it, see many of the news articles cited in this chapter, for starters.

2. For probably the two most comprehensive and definitive Himalayan and Everest mountaineering histories, see, respectively, Isserman and Weaver 2008; Unsworth 2000. For an account of how mountaineering literature reflected mountaineering culture's masculine and imperial ideologies, see Bayers 2003.

3. This is actually a fairly widely accepted truth about the culture of Mount Everest today. Besides the books cited in the note above, see also Heil 2008; Kodas 2008. For perhaps the most well-known propagation of this accepted account, see Krakauer 1997.

4. See, for example, the vast scientific literature on global warming. See also Burke 2010.

5. In addition to Burke 2010, see Gillespie 2010.

6. As I write, news of the youngest climber to reach the summit of Mount Everest, thirteen-year-old Jordan Romero, has been making headlines.

7. For just a couple of examples, see Hall 2007; Weathers 2000.

8. Here I am taking a cue from a broad range of literature theorizing the disgusting. See, for example, Douglas 2002; Miller 1998; Menninghaus 2003. That which is disgusting exhibits qualities such as not belonging, decay or degenera-

tion, hyperfecundity (as with insects or bacteria), and moistness when dryness is expected. The disgusting is a deeply embedded cultural conception, and probably at least partly installed through evolutionary processes, so that the reader's imagination can probably be trusted to have an automatic association with what counts as disgusting.

9. For an extended historical analysis of how modern subjectivity came to be related to the scientific method, see Shapin and Schaffer 1989.

References

Agence France-Presse. 2009. Nepalese Sherpa Scales Qomolangma 19th Time. *China Daily*, May 22. http://www.chinadaily.com.cn//-05//_7927461.htm.

Bayers, Peter L. 2003. *Imperial Ascent: Mountaineering, Masculinity, and Empire.* Boulder: University Press of Colorado.

Bishop, Barry. 1963. How We Climbed Everest. *National Geographic* 124 (October): 477–507.

Bishop, Brent, and Chris Naumann. 1996. Mount Everest: Reclamation of the World's Highest Junk Yard. *Mountain Research and Development* 16 (3): 323–327.

Buncombe, Andrew. 2009. Climb to the Moral High Ground: A Record-Breaking Ascent on Everest Is Also a Bid to Expose Its Degradation. *Independent* (London), April 7. http://www.independent.co.uk/news/world/asia/everest-climb-to-the-moral-high-ground-1664457.html.

Burke, Jason. 2010. Team Sets Out to Clear Bodies from Everest's Death Zone. *Guardian*, April 19. http://www.guardian.co.uk/////everest-death-zone-clean.

Connor, Steve. 2010. Climate Change Will Melt Snows of Kilimanjaro "Within 20 Years." *Independent*, August 15. http://www.independent.co.uk//change/change-will-melt-snows-of-kilimanjaro-within-20-years-1813631.html.

Cronon, William, ed. 1996. *Uncommon Ground: Rethinking the Human Place in Nature.* New York: W. W. Norton and Company.

Douglas, Mary. 2002. *Purity and Danger: An Analysis of the Concepts of Pollution and Taboo.* London: Routledge Classics.

Elmes, Michael, and David Barry. 1999. Deliverance, Denial, and the Death Zone: A Study of Narcissism and Regression in the May 1996 Everest Climbing Disaster. *Journal of Applied Behavioral Science* 35 (2): 163–187.

Gillespie, Ian. 2010. I've a Mountain of Cleaning to Do, Too. *Toronto Sun*, April 24. LexisNexis Academic.

Hall, Lincoln. 2007. *Dead Lucky: Life after Death on Mount Everest.* New York: Tarcher.

Hawkins, Gay. 2006. *The Ethics of Waste.* Lanham, MD: Rowman and Littlefield.

Heil, Nick. 2008. *Dark Summit: The True Story of Everest's Most Controversial Season.* New York: Henry Holt and Company.

Isserman, Maurice, and Stuart Weaver. 2008. *Fallen Giants: A History of Himalayan Mountaineering from the Age of Empire to the Age of Extremes*. New Haven, CT: Yale University Press.

Ives, Jack D. 2004. *Himalayan Perceptions: Environmental Change and the Well-being of Mountain Peoples*. London: Routledge.

Ives, Jack D., and Bruno Messerli. 1989. *The Himalayan Dilemma: Reconciling Development and Conservation*. London: Routledge.

Kodas, Michael. 2008. *High Crimes: The Fate of Everest in an Age of Greed*. New York: Hyperion.

Krakauer, Jon. 1997. *Into Thin Air: A Personal Account of the Mount Everest Disaster*. New York: Villard Books.

Menninghaus, Winfried. 2003. *Disgust: Theory and History of a Strong Sensation*. Albany: State University of New York Press.

Miller, William Ian. 1998. *The Anatomy of Disgust*. Cambridge, MA: Harvard University Press.

Mok, Kimberly. 2010. Everest Expedition to Clean World's Highest Garbage Dump. *Treehugger*, April 19. http://www.treehugger.com////expedition-to-clean-worlds-highest-garbage-dump.php.

Rogers, Clint. 2007. *The Lure of Everest: Getting to the Bottom of Tourism on Top of the World*. Kathmandu: Mandala.

Rogers, Heather. 2010. *Green Gone Wrong: How the Economy Is Undermining the Environmental Revolution*. New York: Scribner.

Shapin, Steven, and Simon Schaffer. 1989. *Leviathan and the Air-Pump: Hobbes, Boyle, and the Experimental Life*. Princeton, NJ: Princeton University Press.

Thompson, Michael, Michael Warburton, and Tom Hatley. 1986. *Uncertainty on a Himalayan Scale*. London: Milton Ash Editions.

Unsworth, Walt. 2000. *Everest: The Mountaineering History*. 3rd ed. Seattle: Mountaineers.

Weathers, Beck. 2000. *Left for Dead: My Journey Home from Everest*. New York: Villard.

Willis, Grania. 2010. Cleaning Up the Earth's Highest Rubbish Tip. *Irish Times*, April 28. LexisNexis Academic.

III

The Cultural Contradictions of Garbage

7

Purification or Profit: Milwaukee and the Contradictions of Sludge

Daniel Schneider

In 1927, the Milwaukee Sewerage Commission took to the airwaves to publicize an "epoch making achievement." "For the first time in the history of sanitation," the radio broadcast claimed, "the valuable plant food elements contained in sewage and trade wastes are being converted into a marketable fertilizer." The Milwaukee fertilizer project was "being watched in all parts of the world," the commission declared (quoted in How Fertilizers Help 1927). This was not hyperbole. Sanitary scientists internationally were paying rapt attention to Milwaukee's experiments in turning sludge into fertilizer, for they revived hopes held since the early days of sewage treatment that sewage could "be converted into gold."[1]

For over sixty years, sanitarians had been pursuing the idea that the nutrients retained in domestic and industrial waste could make sewage treatment pay for itself by recycling sewage sludge into fertilizer. Towns could purify their sewage and make a profit too. Processes were patented, companies were capitalized, and franchises were granted—all with the hope of making money from sewage. All of these plans were failures.[2] By 1913, when Milwaukee established its sewerage commission, it had become clear to many, if not most, sanitarians that profit and purification were inherently contradictory goals. It might be possible to make money from sewage, they concluded, but only at the expense of the goal of sanitation. Conversely, sewage could be purified, but towns must expect to pay for it. As one prominent sanitarian declared, "The two results of profit, or even solvency, and purification, are inconsistent and incompatible with each other" (Bazalgette 1876–1877, 113–114).

It was not that sewage did not contain valuable components. Rather, the expense of separating the valuable from the worthless was greater than the market value of the sewage. "The valuable constituents of sewage are like the gold in the sand of the Rhine; its aggregate value must be immense, but no company has yet succeeded in raising the treasure," stated German chemist A. W. Hoffman in 1857 (quoted in Royal Commission 1884, xxxiii). Now the Milwaukee Sewerage Commission had appeared to finally raise that treasure, and was marketing it over the radio "at a very reasonable price put up in 100 lb. bags" (quoted in How Fertilizers Help 1927).

The Activated Sludge Process

Milwaukee, located on the shores of Lake Michigan, faced a public health crisis in the early twentieth century. The city used the lake as both a water supply source and the ultimate sink for its waste. Milwaukee's drinking water was increasingly contaminated by sewage entering the lake from the three rivers that drained the city, causing repeated outbreaks of typhoid fever. When the Milwaukee Sewerage Commission was established, one of its first tasks was to choose a treatment method to be used in a new sewage treatment plant (Copeland and Wilson n.d.).[3]

Just as the commission was beginning its studies of the Milwaukee situation, a trio of British chemists—Edward Ardern, William Lockett, and Gilbert Fowler—unveiled a new sewage treatment process they called activated sludge. In this process, sewage was placed in tanks and bubbled with air. Large populations of bacteria grew in the roiling tanks, feeding on the sewage. After treating the sewage for just a few hours, the bacteria-rich sludge was allowed to settle out, cleansing the sewage of solids and leaving a clear effluent. Like a sourdough starter, some of the settled sludge was added back into the aeration tank to treat the new sewage. The engineering press quickly heralded the activated sludge process, and cities around the world began to experiment with it.

The remaining sludge, though, required disposal, and this disposal represented one of the greatest problems for sewage treatment. The first chemical analyses of activated sludge revealed a high nitrogen content—almost twice that of the sludge from other processes. With these analyses, the hope of profitable utilization of sewage was revived. Fowler entertained such high expectations of the profitability of activated sludge that he proclaimed, "The fertiliser end of the process will come to be seen as the most important, sewage purification being, as it were, a by-product" (Fowler to Bartow 1915). With activated sludge, "we have the master key with which to unlock the treasure stored up in sewage sludge," wrote an editor of the *Engineering Record* (Commercial Possibilities 1916).

Milwaukee, whose sewage commission's chief engineer was T. Chalkley Hatton, became the first large city to initiate a sewage treatment plant based on the activated sludge process. From the beginning, the manufacture of a salable fertilizer was an integral part of its sewage treatment plans. As with every city, Milwaukee's decision to build sewage treatment along with the form that treatment would take was a complex combination of politics, available technology, economics, expediency, and the actual condition of the rivers, lake, and water supply.[4] The political pressures in the city to deal with its sewage problem fortuitously coincided with the availability of the activated sludge process, making that technology a possible response. Further, because activated sludge required less land, the sewage plant could be built on a relatively small parcel near the harbor—a location that would not exacerbate political and ethnic tensions between the North and South Side wards of the city. Finally, the promise of revenues from fertilizer helped overcome objections to the relatively high initial capital costs.

But the choice was not overdetermined. Milwaukee's socialist city government and the Milwaukee Sewerage Commission disagreed on the importance of sewage treatment versus water filtration, with city officials strongly favoring water treatment and cheaper sewage alternatives. The activated sludge process was by no means an established technology, and it took faith on the part of Hatton and the commission to pursue it. Because of Milwaukee's history of municipal utilities and the powerful political movement that put municipal trading at the center of its agenda, Milwaukee's decision to incorporate the sale of fertilizer as an integral part of its plans was a powerful argument in favor of sewage treatment over water filtration. The commission's chemist, William Copeland, estimated that fertilizer from activated sludge would cost from eight to twelve dollars a ton to produce, but could be sold for nine to fifteen dollars per ton. "The recovery of nitrogen in sewage sludge has at last been brought within the range of a commercially practicable problem," he concluded (Nitrogen Recovery 1916, 445).

The estimate of nine to fifteen dollars a ton, however, was entirely theoretical because there was no existing market for sewage sludge. As another activated sludge researcher had earlier commented, "It is useless to calculate the exact value of activated sludge since a market must first be established." Yet marketing was not one of the roles that sanitary engineers typically considered part of their purview or one in which they had much expertise. "It is not customary," Hatton remarked, "for an engineering organization engaged in constructing a plant to be interested in the work of marketing that which is to be produced" (quoted in Eleventh Annual Report of the Sewerage Commission 1925, 80).

Studies funded by the sewerage commission showed that activated sludge was effective as a component of mixed fertilizers. Armed with these results, the commission set out to market the sludge to fertilizer manufacturers. For manufacturers to use sludge, though, they had to overcome a powerful prejudice against sewage materials. For instance, one large manufacturer refused to prepare the mixtures for experimental plots after laborers in the plant refused to work with it "because the peculiar odor made them sick." Partly as a result of these experiences, advertising for Milwaukee's activated sludge downplayed its origins in sewage and emphasized its "earthy" odor, while reiterating that it was a "sterile manufactured product." But until that marketing was successful, the commission was warned that manufacturers would not recognize "the difference between activated sludge and raw sewage" (quoted in Noer 1927).

To help overcome this prejudice, one final detail needed to be completed before Milwaukee was ready to market its sludge. To attract the trade from fertilizer manufacturers, golf courses, and flower growers, the markets identified as the most promising, "it appeared necessary to adopt a 'trade name,'" reported Hatton. The engineer felt that it was critical that the name should not "suggest sewage sludge" but rather "its ingredients appropriate for plant food." The commission sponsored a contest for naming its sludge, and advertised in national fertilizer journals. A fertilizer dealer in South Carolina submitted the winning entry, coining the name Milorganite, for Milwaukee organic nitrogen (quoted in Twelfth Annual Report of the Sewerage Commission 1926, 9).

The sewerage commission was new to marketing, and its early efforts showed a poor grasp of how to market a product to a specific industry. The commission asked Alexander M. McIver and Son, the South Carolina fertilizer dealer that had named Milorganite, to provide a private critique of its first marketing materials targeted at the fertilizer manufacturers. In response, the firm cautioned Milwaukee to remember its intended market. The firm saw a booklet heavy on data and the kinds of evidence that had been appearing in the scientific literature, and lacking an appreciation for how fertilizer manufacturers would respond, McIver and Son urged particular caution about Milwaukee's claim that in contrast to Milorganite, high-analysis fertilizers might injure crops. As the fertilizer industry had "been doing everything possible to raise the percentages in their mixed fertilizers," McIver and Son thought a booklet making contrary claims "might not appeal to them, especially if they thought a copy of this booklet might accidentally get in the hands of the farmers." Further, the sales manager, V. H. Kadish, had included photos and data in the booklet showing that yields on certain crops were comparable between Milorganite alone and chemical ammonia. McIver and Son pointed out that these data would appear to the manufacturers as critical of their products. Rather than encouraging Milorganite's use, these results would alienate its intended market. Instead of comparing Milorganite to chemical nitrogen, Milwaukee should stress the comparison between Milorganite and other organic nitrogen sources, particularly packinghouse tankage. Finally, McIver and Son recommended that the signs in the photos that labeled which fertilizers were used should somehow be changed to improve the advertising copy. Produced before the name Milorganite had been coined, the signs simply said "sludge" (Alexander M. McIver and Son to V. H. Kadish 1926).

Milwaukee began commercial production of Milorganite in 1926. In the first years of production, shipments of Milorganite to fertilizer manufacturers dominated the sludge sales. But at the same time, Milwaukee started to develop a niche market for Milorganite in sales to golf courses. It turned out that the chemical composition of activated sludge was ideal for promoting turfgrass growth. Milorganite possessed other advantages as well compared to the products then used on turf. Because the sludge was heat dried, weed seeds were killed. Many greenkeepers liked to avoid using potash in their fertilizers because potash promoted clover growth. Unlike manure, activated sludge contained negligible potash. What was a disadvantage in the mixed fertilizer market was an advantage in the specialty golf market. Finally, the potential profit was much higher in sales to golf courses, because activated sludge can be used straight instead of one ingredient in a complete mixed fertilizer. Milwaukee was able to sell to golf courses at a substantial premium over sales to fertilizer manufacturers (Noer 1925; Tentative Program 1924).

Milwaukee began to emphasize the use of Milorganite for golf courses and developed an infrastructure for marketing the fertilizer to this specialty market. The Milwaukee Sewerage Commission essentially established an agricultural extension service for golf courses. Named the Turf Service Bureau, it provided a number of services to golf courses, including soil testing. Milwaukee ran into problems in marketing Milorganite, and quickly found that maintaining fertilizer sales required an enormous amount of institutional effort, blurring the commission's roles in purifying sewage and operating a fertilizer factory. Soon after Milorganite's introduction in 1926, the stock market crash and Great Depression ate into the market for both the mixed fertilizer and particularly specialty sales to golf courses. Golf clubs significantly reduced their spending on upkeep as they lost members to the Depression. Milwaukee was competing against chemical fertilizer manufacturers as well as companies like Armour and Swift that were marketing their meatpacking by-products as fertilizer. To counter these trends, Milwaukee continued to advertise extensively, with ads appearing in almost every issue of the magazines devoted to golf turf management.

Additionally, to counter the reduced spending on maintenance by golf courses, Milwaukee produced advertisements to convince golf course managers that the state of the turf could help attract and keep members. "Every Greenkeeper Now a Committee of One on Membership" proclaimed one ad. "It's largely your responsibility . . . this holding of memberships. And *good fairways* is the answer," the ad continued. "Oh the Crime of it! letting him starve," another ad noted, complete with pictures of "Mr. Starving Turf," which might have resonated with reader's images of breadlines. "Milorganite's Big Economy Plan," the ad declared, "*permits* even the most distressed clubs to continue a sound improvement program." In an economic environment where even the best clubs were under financial strain, the city tried to sell fertilizer by convincing smaller courses that attention to their greens could leapfrog them over their competitors. Milorganite "*creates* new Leaders out of Secondary Courses," ads claimed, and the commission enjoined, "Milorganize your course" (Advertisement 1931). By the end of 1932, it was apparent that the commission's marketing plan was successful when it declared, "More Golf Clubs use Milorganite than any other Fertilizer (advertisement 1932a, 1932b, 1932c).

Milorganite dominated, but it dominated a shrinking market. With the fertilizer market depressed, the commission officials considered abandoning the production of Milorganite and incinerating the waste sludge instead. Chief Engineer Townsend was adamantly opposed, however, and produced an economic analysis that showed if Milorganite could be marketed at eight dollars per ton, it would be preferable to incineration. Given that the average sale price of Milorganite in the previous three years was over sixteen dollars a ton, Townsend concluded that abandoning Milorganite production appeared "both illogical of thought and unwarranted." Yet this conclusion depended completely on Milwaukee's continued success in marketing the material (quoted in Nineteenth Annual Report of the Sewerage Commission 1932, 172).

As opposed to switching to incineration, Milwaukee doubled down on Milorganite and initiated marketing plans that prefigured some of the vast changes about to take place in the agricultural chemical industry as a whole. Like others in the agricultural industry, the city sought to diversify its product line to increase sales. The granular nature and low water content of Milorganite made it an excellent carrier for other agricultural chemicals, and Milwaukee agronomists had encouraged golf course managers to mix Milorganite with lead arsenate (an insecticide in common use) as well as other chemicals used in weed and disease control.

The success of Milorganite as a carrier led the sewerage commission in 1940 to begin marketing its own specially formulated combination of fertilizer and pesticide. The sewage plant combined Milorganite and sodium arsenite, a weed killer, premixed at the Milorganite production facility rather than by customers. The commission at first described the product in advertisements simply as a "Milorganite-Arsenite mixture," but quickly ramped up its marketing, coining a new name, "Milarsenite," and producing special bags, with Milorganite's trademark symbol adapted for the new product. The arsenite in the mixture killed the weeds while the fertilizer induced the "grass to spread and fill the voids left by the weeds." By combining the two products in one formulation, the commission pioneered the production of what later would become known as "weed-and-feed" applications. By the 1990s, weed-and-feed products dominated the home fertilizer market, with more households buying them than all other product types combined (National Gardening Association 1996, 87; 1991, 86). The development of fertilizers premixed with herbicides has been associated with Scotts' first marketing of "Weed and Feed" in 1947. But the development of Milarsenite preceded this by seven years.

In its tests, the sewerage commission found the weed control properties of Milarsenite to be "astonishing" and "almost perfect," with only faint or no discoloration of the grass.[5] Milarsenite was introduced in 1940 in the *Greenkeepers Reporter*, both in an advertisement and announcement. The advertisement, under the slogan "kills weeds, saves grass," reported the results of Milarsenite's use for killing clover and other weeds, and proclaimed "because results this fall were so promising, a limited amount of Milarsenite will be offered again in 1941. If you have a weed problem it will pay you to try Milarsenite next spring." Milwaukee produced a leaflet instructing greenkeepers on the use of Milarsenite and warning that it was poisonous, although "no worse than mercury." The commission recommended that workers wear coveralls, waterproof gloves, and a mask, or at least a "pad of folded cheesecloth worn over the mouth and nose." While filling the spreader, workers should "stand on the windward side." Photographs of workers applying the product showed that they did not heed these warnings, though (How to Use MILARSENITE n.d.).

Milwaukee began to look for other potential products derived from Milorganite that could become the basis for marketing campaigns. Activated sludge, and thus its product Milorganite, was essentially composed not of sewage but rather of the cells of the bacteria that grew on the sewage. As a bacterial biomass, it was made up of proteins, fats, amino acids, and other organic materials that might have a variety of industrial applications. Other waste products, like tankage from slaughterhouses, had proved to be good sources of not only the proverbial horsehide glue but also other adhesives, foaming agents, protein supplements, and the like. Milwaukee commissioned an industrial research company, the Miner Laboratories of Chicago, to explore the potential of Milorganite "as an industrial raw material" (Report 44–2 to the Sewerage Commission 1994). The Miner Laboratories had made its reputation by developing a process for extracting furfural, a feed supplement, from waste oat hulls for the Quaker Oats company (Research 1941, 73–74). In its work for the sewerage commission, the lab was to conduct similar research on activated sludge. Miner Laboratories used a variety of different solvents to extract various fractions from Milorganite, and sent these fractions for evaluation to industrial companies throughout the United States, like the US Gypsum Company, International Minerals and Chemical Corporation, and Commercial Solvents Corporation. One of Miner's first discoveries was that various extracts of Milorganite could provide a previously unknown growth factor for yeast and bacteria that accelerated a variety of industrial fermentations, including ethanol, bread dough, lactic acid, and acetone-butanol.

Marketing products from sewage sludge, however, continued to face deep-seated antipathy from users toward including human waste in products for human consumption. The Milorganite yeast activator was used in beverage alcohol as well as industrial alcohol fermentations. On learning of this, the Food and Drug Administration (FDA) issued a stop order, declaring "the idea of using such a substance in the production of alcohol for beverage purposes is so repugnant that the proposal should not be considered" (C. W. Crawford to O. J. Noer 1951). While untroubled by the use of Milorgante for industrial alcohol, the FDA threatened to take action against any use of the fermentation activator for beverages.

While activated sludge could activate commercial fermentations, its mode of action was unknown. Miner Laboratories suspected that Milorganite provided a previously unknown B-complex vitamin. In searching for the specific activation factor, it discovered that activated sludge "was one of the richest sources yet discovered for Vitamin B_{12}," which had only recently been discovered and identified as a key animal nutrient in 1948.[6] The presence of vitamin B_{12} in activated sludge suggested additional uses for Milwaukee's sludge as an additive to animal feeds, and became a focus of Miner Laboratories' work. Given the FDA's action against Milorganite, the commission was surprised when the following year, the FDA indicated it might be willing to approve Milorganite as a feed supplement based on its vitamin B_{12} content. Nevertheless, wary of the public perception of such a supplement, the commission sought the advice of the major feed producers in the United States on their willingness to consider Milorganite as a vitamin B_{12} supplement in animal feed. Research directors for Quaker Oats, Allied Mills, and the Ralston-Purina Company all agreed that regardless of the federal approval of Milorganite as a supplement, they would be unwilling to use it directly in their feeds. There had been a recent scare when bonemeal used in feeds had caused an outbreak of anthrax in the Midwest, and feed mixers were becoming increasingly cautious in using suspect ingredients. If they used Milorganite, they feared being "accused of using a sewage product." But these concerns applied only to using bulk Milorganite directly as a feed supplement. These advisers suggested that the industry would be quite willing to use an extract of Milorganite in which the vitamin had been concentrated, provided it had "a proper name" and the "source of the extract would not have to be given" (Report on the Attitude of Three Large Companies 1952).

Producing Milorganite was a fundamental part of the disposal process, providing a way to dispose of the sewage. But commissioners did not see vitamin B_{12} extraction as necessary to the treatment process. Rather, it represented "a new commercial undertaking" for which they felt it was inappropriate to expend public funds. The commission thus sought a private company to build and own the plant, with Milwaukee supplying the space. In return, Milwaukee would get 35 percent of the net profits (Vitamin B_{12} Test Project 1953).

The sewerage commission contracted with the Verne E. Alden Company, a Chicago engineering firm, to build a vitamin B_{12} extraction plant at Jones Island, selling the vitamin for use in poultry and animal feeds. It hoped to produce vitamin B_{12} for humans as well, making history as the first "by-product from sewage for human consumption" (B_{12} Vitamin Production 1956, 477). After several delays, in 1956 the Alden Company finally built a pilot extraction plant in Milwaukee's fertilizer warehouse. When anticipated orders for the extract failed to materialize, though, the plant closed temporarily in 1958. Restarted in 1959, it closed permanently after only a few months in operation (Forty-Third Annual Report

1957, 33; Forty-Fifth Annual Report 1959, 31; Forty-Seventh Annual Report 1961, 32–33).

Despite the hopes that "Milorganite may well turn out to be another coal tar in the biochemical field," none of the industrial applications for it panned out (Benefits Seen in Milorganite 1952). Milorganite continued to be a successful fertilizer for both the home and golf course markets, however. Seeing its success, several other cities entered the market. Pasadena started selling its sludge in 1928, Houston followed suit in 1932, and Chicago did the same in 1939 (Utilization of Sewage Sludge 1946, 58). Whereas Milorganite became a well-known brand, none of these cities was able to develop a consistent marketing plan. Houston's market was restricted by the limited production of fertilizer material. The Houston treatment plant produced only about two thousand tons per year of activated sludge, compared with fifty thousand tons in Milwaukee. Similarly, Pasadena, despite the production of extensive marketing materials, did not stay in the market long, abandoning its fertilizer production in 1942. Chicago's lack of a consistent name for its fertilizer suggests that marketing was much less of a priority than in Milwaukee. Chicago had no marketing staff. Rather, it sold its entire production to one company, which contracted for the sludge. In turn, the contractor assumed "all expense for selling, promotion and credit." Most of the sludge was resold to fertilizer manufacturers in the southeastern states for use on tobacco and cotton (Annual Budget Report 1940). Receipts for Chicago's sludge grew throughout the 1940s, until 1948, when Chicago sold over one million dollars' worth of sludge (Report to the Board of Trustees 1949).

The success of cities in marketing their sludge depended on a number of factors. The most salable type was the heat-dried activated sludge produced by Milwaukee, Pasadena, Houston, and Chicago. This sludge had the highest concentrations of nitrogen and lowest water content, making it economical to transport long distances. Of lower value was digested activated sludge, which had lost much of its nitrogen content through the digestion process. Finally, of even lower value was the settled sludge from other treatment processes like Imhoff tanks. The equipment required for heat drying was expensive, and thought to be economical only for the largest cities. Experts considered it impossible for small cities producing Imhoff sludge to sell it as fertilizer. They could successfully dispose of their sludge to local agriculture, but it would have to be given away and loaded for free. Yet marketing was essential even for giving away the sludge. P. N. Daniels (1934, 277, 279), superintendent of the Trenton plant, noted that of the three hundred plants in New Jersey, "the sludge cannot be given away, not to mention selling it."

What most cities agreed on was the need for a catchy name for their product. "A fancy name such as 'Super-humus,' 'Grozall,' or 'Fortified humus,' helps a lot," observed Daniels (ibid., 81). Most names incorporated some version of the city name in conjunction with a phrase suggesting growth or naturalness. Thus, Toledo produced "Tol-e-gro, " and Akron made "Akra-Soilite." Perhaps the most unfortunate of these names came from Clearwater, Florida, which called its product "Clear-o-sludge" (Utilization of Sewage Sludge 1937, 903; Collings 1955, 131). "Hou-Actinite" (presumably for Houston Activated Nitrogen) was a play on Milwaukee's trademarked sludge. "One of the tricks in advertising," according to Los Angeles' plant manager, involved writing the name of its fertilizer, "Nitrohumus," by sprinkling fertilizer on a potential customer's lawn in the shape of the letters. "In a week or ten days," he recounted, "the name of the sludge is written in bright green across the yard" (Rawn 1941, 187).

To protect its own investment in developing the trademark Milorganite, Milwaukee tried to prevent other cities from co-opting its marketing by developing similar names. Milwaukee sued a number of cities for adding the suffix "-organite" to their product names. The Kelly Agricultural Products Company, located near Pittsburgh, had named a fertilizer derived from waste leather "Pittorganite." Milwaukee took the company to court, claiming a trademark on any name combining the phrase -organite, and forced it to change its name (it apparently chose "Groganic"). While the settlement netted the commission only twenty-five hundred dollars, the lawyers pointed out that having its trademark validated in court was worth many times that amount (Forty-First Annual Report 1955, 20; Fortieth Annual Report 1954, 28; Forty-Second Annual Report 1956, 13–14).[7]

Despite the conclusion that heat drying would be economical for only large cities, several smaller cities tried to make a go of the fertilizer business. Grand Rapids, Michigan, started its sewage plant with high hopes of turning a profit. "Efficient utilization by the city of by-products produced at the new sewage disposal plant may make the enterprise an industry which is virtually self-sustaining," claimed the city service director. But before the plant could turn a profit, the director admitted that he had to "seek a market." Grand Rapids marketed its sludge as "Rapidgro," and began its fertilizer production in 1931, in the midst of the Depression. To get the project started, Superintendent James R. Rumsey used a variety of strategies for building the plant and marketing the fertilizer. He utilized scrip labor to build the fertilizer plant. Workers "from the city's unemployed were paid in scrip which they exchanged at the city store for food or clothing." With no market initially, Rumsey turned the city into its own customer. Various city departments used all of the production. The sewage plant used the fertilizer and scrip labor to beautify its grounds. It also used the fertilizer to build up "wasteland" and turned it into a city garden that provided vegetables to the city's welfare department, supplying food for the indigent. It also enlisted county farm agents to advise local farmers on the use of Rapidgro, and distributed samples to local garden clubs and fruit growers. Finally, the sewage plant, following Pasadena's example, enlisted unemployed men to act as commissioned salespeople for Rapidgro.[8]

Selling mostly to a limited local market, Grand Rapids nevertheless managed to sell a thousand tons of sludge per year, earning a net profit on fertilizer sales of between $1,000 and $10,000 annually. Yet the city was embarrassed when it became known that the parks department, rather than using its own Rapidgro, had ordered eight tons of Milorganite for use on the city's parks and cemeteries. When asked by a furious city council member to explain his order for Milorganite, the parks superintendent said, "The Milwaukee product has a faster action than the fertilizer we make at the sewage disposal plant." "To save face," the city announced that it would enrich Rapidgro with extra nitrogen for use by the parks department. "It will cost an extra $10 a ton to bring the local product up to the potency of some of the nationally-distributed brands," stated the *Grand Rapids Herald*.

By the 1950s and into the 1960s, most producers of fertilizer material from sludge left the business. Finding markets drying up and production more trouble than it was worth, city after city abandoned its sludge recycling program, opting to either incinerate or landfill its sludge (Caster 1955). Despite the continued albeit modest success of Rapidgro, for instance, Grand Rapids abandoned the manufacture of fertilizer in 1966 and began taking all of its sludge to the city dump. Chicago found that no companies would bid for its sludge. With no outlet for its production, it briefly looked into packaging its fertilizer and selling it directly, as Milwaukee did, but these plans never materialized (Olis Proposes Package Sale 1954). Instead, Chicago started dumping its fertilizer on the banks of the Sanitary and Ship Canal, and encouraged the public to remove it free of charge (Sanitary Unit 1957). In 1965, Chicago stopped making fertilizer, and planned to demolish the plant in 1969 (Sanitary District 1968). By 1960, when the Water Pollution Control Federation updated its 1946 manual of practice on the utilization of sewage sludge, it concluded that "the sale of sludge can never be expected to provide a net return over processing costs" (*Utilization of Municipal Wastewater Sludge* 1971, 40). Selling sludge might be less costly than other disposal methods, but the hope of "unlock[ing] the treasure stored up in sewage sludge" was not realized. As English firms had discovered in the previous century, profit and purification remained as contradictory as ever.

The relentless pursuit of income from sludge also posed fundamental problems for a plant's ability to treat its sewage. The contradiction between purification and profit was real. As English plants had discovered in the nineteenth century that "a profit may be made without purification, and very frequently the purification may be effected without profit, but the two cannot apparently be combined," so Milwaukee faced the explicit trade-off between selling Milorganite and treating its sewage. The capacity of the Milorganite plant to produce fertilizer from sewage was smaller than the capacity of the activated sludge tanks to treat the sewage. During large storms, as increased volumes of sewage made its way to the plant, Milwaukee was forced to choose between running the plant at the capacity of the activated sludge tanks or Milorganite production process, allowing the excess sewage to bypass the plant and go into the lake untreated.

James Brower, plant superintendent from the 1930s to the 1950s, recounted how he thought the aims of the sewage treatment plant were being subverted by the need to market fertilizer. "I have always believed and still do, that the plant was designed and built to treat sewage, that the main purpose was to keep the rivers and lake as free from sewage polution [*sic*] as was humanely possible, regardless of what happened to the fertilizer." Nevertheless, concerns over marketing caused the commission to release untreated sewage into the lake during storms even though the plant had the capacity to treat it. The commission made the explicit decision to sacrifice treatment to the limitations of fertilizer production. "This was mainly done," Brower charged "to maintain a high percentage of nitrogen so that the sales department could meet the guarantees as printed on the bags."

Dilution of the sewage by large rainstorms would decrease the percentage of nitrogen in the fertilizer. Even though the health department and water purification plant complained to the sewerage commission, James Ferebee, the chief engineer at the time, refused to alter the practice. Concerned about the pollution, Brower demanded a written order from Ferebee before he would bypass the plant. In 1941, Ferebee issued that order, and declared that "until further notice, the quantity of sewage taken in by the purification plant shall be reduced to the lowest possible operating quantity during times of quick, intense rain storms. The purpose of this order is to decrease the dilution of organic material." After continued complaints, though, Brower was able to convince Ferebee to utilize the purification plant's full capacity during storms. That decision was quickly reversed, and Brower claimed that "someone got to him, stating that after a storm it took a considerable amount of time to bring the fertilizer back to the standard guarantee." (Brower n.d., 37–38) The drive for profit had distorted the original goal of the plant to treat sewage and protect public health.

In response to these structural contradictions between profit and purification, Milwaukee continued to produce Milorganite, but the sewerage commission's perspective began to slowly change. Over time, the commission thought of itself less and less as a producer of fertilizer. When the commission had to build a new plant to accommodate growth on the south side of the city, fertilizer production became a subsidiary goal. Half a century earlier, the potential to profit from its sewage had lured the Milwaukee Sewerage Commission and its engineer Hatton to construct a fertilizer plant as an integral part of its sewage treatment facility. Now, in a newspaper article headlined "Milorganite Moves from Profit to Loss," plant officials regretted the decision: "If not saddled by the 10 million dollars tied up in the plant necessary to produce Milorganite, it would be better for the district to burn the waste" (Milorganite Moves from Profit to Loss 1960).

When the commission built the new south side treatment plant in 1968, it decided it would be cheaper to digest its activated sludge and incinerate it than to produce Milorganite (S. Shore Sewage Unit 1956). But the contradictions of sludge persisted. During a 1972 meeting with some of Milorganite's largest customers, sales manager Charlie Wilson painted a rosy picture. Two years after the first Earth Day and barely a month after passage of the Clean Water Act, the Milwaukee Sewerage District was selling every ton of Milorganite it produced, and Wilson thought if the current "ecology kick" continued it could probably sell twice that. The chief concerns were expanding production and maintaining the ability to keep profit margins "undoubtedly the highest in the industry" (Milorganite Advisory Council Meeting 1972). There were some clouds on the horizon, though—ironically from the very same ecology kick that Wilson hoped would spur sales.

By the middle of the decade, Milwaukee faced a deepening crisis in disposing of its activated sludge. Having held a virtual monopoly in marketing sludge for over forty years, the commission was suddenly faced with increasing competition from other cities that, because of the provisions of the 1972 Clean Water Act and other legislation, began marketing their sludge as well. A new batch of sludge products appeared, such as "Metrogro," "Biogro," and "Gardenlife" (Flynn 1982). In 1988, the federal government instituted a ban on the dumping of sludge in the ocean. Major East Coast cities like Boston and New York that had been dumping their sludge were forced to find other outlets. Rather than dumping the sludge at sea, these cities began dumping it on the market, eating into Milwaukee's market share and profit (Rockwell 1993; Barron 1992).

In addition, there was increasing concern over the safety of sewage sludge as a fertilizer, reducing the ability of any city to sell its sludge. There were a number of scares in the 1970s and 1980s involving the safety of Milorganite, resulting from a fundamental contradiction of sewage treatment. Sewage contained both industrial and municipal waste, including heavy metals and other contaminants. The more dissolved and particulate material that was removed by the sewage treatment process, the cleaner the effluent from the plant, and the less pollution of rivers, lakes, and ocean. But it also meant a greater concentration of these pollutants in the resulting sludge. As long as nutrients were the main component of sewage, this led to a more marketable sludge. When the widespread use of phosphates in detergent led to an increase of phosphorus in the sludge, for instance, Milwaukee trumpeted the higher concentrations as an improvement in the fertilizer. Yet when sewage contained toxic waste—either from industrial discharges, or household waste like solvents, paints, or other material washed down the sewers—the cleaner the effluent, the more toxic the sludge. Increasing the purity of the wastewater ate into the profits from selling the sludge.

In 1974, recognizing the potential for sludge to contain high concentrations of pollutants, the EPA began to develop regulations limiting heavy metals in sludge, much to the consternation of many municipal sanitary districts. The EPA's proposed regulations would have excluded from use over 90 percent of the municipal sludge from Wisconsin and Indiana, and 80 percent from Illinois. Individual states also started to develop regulations. Following the EPA's example, Wisconsin published guidelines in 1975 on toxics in sludge (Vinton W. Bacon to Robert J. Borchardt 1977). The sewerage district, fearing the effects on Milorganite, complained to the state about rigid limitations on heavy metals (Robert J. Borchardt to David G. Berger 1977). At the same time, it approached members of its Milorganite Advisory Council—made up of fertilizer distributors and other large users of Milorganite—to contact their congress members to help prevent the US Department of Agriculture and EPA from "endangering" Milorganite with demands for health warnings (Milorganite Advisory Council Meeting 1976).

With municipalities in Wisconsin reluctant to address the issue of heavy metals, a Milwaukee environmental organization, Citizens for a Better Environment, began a campaign against the use of sewage sludge contaminated by heavy metals. In 1978, the organization issued a well-publicized report on the high cadmium levels in Milorganite (Milorganite Gardening 1978).[9] Milwaukee, to its credit, responded to the publicity by placing a prominent label on its Milorganite bags, warning in bold lettering: "DO NOT USE ON VEGETABLE GARDENS, OTHER EDIBLE CROPS OR FRUIT TREES. Eating food grown on soil containing Milorganite may cause damage to health." While the warning was limited to Milorganite's use on edible crops, consumers understandably began to shy away from it for all uses. "Even though the notice pertains to home grown vegetable production, many homeowners do not want their children or even family pets walking on lawns that receive Milorganite 'heavy-metal cadmium' applications," stated the marketing staff (Swanson to Hawkins 1989). The warning label caused an almost immediate 30 percent drop in bag sales (Resolution Authorizing Approval 1993).

In 1987, the health effects of Milorganite were further questioned when a former quarterback for the San Francisco 49ers was diagnosed with amyotrophic lateral sclerosis (ALS), or Lou Gehrig's disease. At the same time, it was revealed that he was only the most recent member of his 1964 team to contract the disease; two others had already died. Public health officials speculated that this unusual cluster might be due to the heavy metals in Milorganite, which had been used on the 49ers' practice field. A Milwaukee physician then made news by claiming that an unusual number of his ALS patients had used Milorganite as well. Researchers investigating the link between Milorganite and ALS found no evidence that the fertilizer was responsible for the disease. Yet as the Milorganite marketing staff recognized, the cadmium warning label raised the question among users, "If Milorganite does not cause ALS, why do you have the warning on the bag?" (Swanson to Hawkins 1989).[10]

To reduce the cadmium content in Milorganite, Milwaukee had passed an ordinance in 1980 requiring the pretreatment of industrial waste. As a result, the cadmium concentrations in Milorganite were more than halved within a couple years. As the sewerage district reduced the concentration of cadmium and other metals in Milorganite, it received enormous pressure from the Milorganite marketing staff to remove the warning label.

Neither the Wisconsin nor federal government had ever required the label, as the EPA had delayed its rules regulating toxics in sewage sludge. Rather, it had been placed voluntarily, albeit under pressure from environmental organizations. Because Milwaukee was competing against other fertilizers that had no label, even though they might have had equivalent cadmium concentrations, the marketing staff, on the advice of its large customers, repeatedly sought to downplay the warnings.

In 1982, with cadmium concentrations in Milorganite further reduced from around 120 to 45 parts per million, the staff convinced the board to downgrade the "warning" to a "notice," and move it from the front of the bag to the back. The district also sought the approval of Citizens for a Better Environment for the label change. With its approval, the label was changed to "NOTICE: Milorganite should not be used on vegetables due to possible uptake of the heavy metal cadmium. When home grown vegetables make up a large portion of the annual diet over a period of many years, cadmium may accumulate in the kidneys, resulting in possible dysfunction." "Safer Milorganite to be sold in snazzier bags," crowed the *Milwaukee Journal* (Zahn 1982b).

Nevertheless, the cadmium content remained a problem; it was too high for Milorganite to be exported to Canada, and other states sought to ban its distribution within their borders (Zahn 1982a). Cadmium concentration continued to decline, though, as Milwaukee enforced its pretreatment regulations. By 1990, concentrations of cadmium were around 5 parts per million, and the marketing staff lobbied to get the label changed once again:

While Milorganite meets all federal and state regulations for fertilizer products, it should not be used on food crops. Use only in accordance with label directions on non-edible plants and grasses. Most natural organic and chemical fertilizers contain trace amounts of heavy metals. Between application, store in a dry area out of the reach of children and away from pets. Wash hands after use. Do not eat.

With increasing concern over the safety of sewage sludge, the EPA released its long-delayed regulations governing the application and disposal of sludge. Mandated by the 1987 amendments to the Clean Water Act, the regulations were finally promulgated in 1993. The regulations divided sludge into several categories based on the concentration of pathogens and ten heavy metals, including cadmium. If the concentrations were below a critical value, the sludge (now termed "biosolids") could be considered of "exceptional quality," and distributed essentially without further regulation. These regulations provided incentives for sewerage districts that marketed their sludge, like Milwaukee, to ensure that their sludge met these criteria. It also helped ensure a market for the sludge by offering reassurance that sludge was, indeed, safe.[11]

With Milorganite's classification as an exceptional quality sludge, the district sought to remove the label altogether. Milwaukee, with the encouragement of the EPA, removed any reference to potential health hazards, stating simply that Milorganite meets "'exceptional quality' guidelines as defined by the Federal Environmental Protection Agency" (quoted in Resolution Authorizing Approval 1993). The EPA regulations had performed their function, as defined by the EPA: encourage the disposal of sewage sludge on land as well as a reduction in toxics concentration. Whether the regulations were sufficiently stringent to protect public and ecological health is still a source of controversy.

Yet the market for Milorganite remained difficult. Even with the changes in the label, dealers reported that homeowners were "still concerned about the metals content" of Milorganite, causing the loss of 60 percent of one dealer's market. Other dealers reported they had stopped carrying Milorganite after the ALS scare and had not carried it since (Market Probe n.d.). Many in marketing felt that the sewerage commission had ignored the importance of generating revenue through Milorganite sales. Rather, they charged, the commission was focusing solely on the quality of the sewage treatment plant effluent. Sewage treatment was caught once again in the contradiction between profit and purification. Wilson claimed that the commission was ignoring Milorganite along with its marketing. "You have said that we are not in the fertilizer business," he wrote to William Katz, a commission director. But "70,000 plus tons of yearly sales and a 4 million plus income," he continued, "could be considered good business by most folks" (Charles G. Wilson to William Katz 1977).

With increasing competition and the warning label depressing sales, the Milorganite division of the sewerage district saw a critical need to improve the marketing of Milorganite, both by creating new products and finding new markets. A 1986 marketing report renamed the contradiction between profit and purification as one of customer versus operations orientation. In the early years of the activated sludge process, the report argued, the sewerage district operated like a "fertilizer factory" focusing on the needs of the fertilizer customer. Recently, however, it had shifted its approach to one of "operations," or purifying the effluent. The report urged that the commission move back toward a marketing orientation (Falvey and Hurtado n.d., 14, 27). In 1991, Milorganite staff members proposed a reorganization that would strengthen the marketing division and allow them more flexibility to create new products. In response, commissioners again raised the contradiction between profit and purification as a fundamental problem in the sewage plant's operation. "Isn't there the potential for conflict between one side of the MMSD which advocates reducing toxics at the source and the other side of the organization which hopes to maximize revenue?" (Ward 1991). Some staff members even wondered if "the MMSD should be engaged in 'business' at all" (quoted in Milorganite Market Positioning Study 1994/1995).

Perhaps, thought the commission, a resolution to this contradiction could be found in spinning off that side of the sewerage district that "hopes to maximize revenue." When the district began vitamin B_{12} production, for instance, these kinds of considerations over the district's proper role in business ventures led it to structure the vitamin production plant as a private operation under contract to the district. By the early 1990s, officials began to investigate the privatization of Milorganite marketing. Houston, after decades of ineffective marketing of its sludge, had started privatizing its sludge operation in 1990. Tampa, New York, Boston, and other cities were about to follow. Houston had contracted with a relatively new private firm, Enviro-gro, to take over its sludge marketing. With EPA regulations imminent, cities were desperate to find ways to dispose of their sludge. Enviro-gro sought to take advantage of this new municipal market, and soon had contracts with Houston, and was in negotiations with both New York and Boston to build and operate sludge drying and fertilizer production as well as market the resulting sludge. Recalling the discussion of sewage from nineteenth-century England, the environmental industry press asked, "Can Cities Turn Profits from Sludge?" (Barron 1992). In answer, an Enviro-gro official pointed to Milwaukee's experience. "Milwaukee has been doing this for so long—and has proven that it is workable—that they have really opened up the marketplace to sludge-based fertilizer." In turn, Milwaukee was looking to the Enviro-gro model to investigate whether it should privatize its own sludge operations.

The marketing staff opposed the privatization of its functions. What Milorganite needed was not a private operator and marketer but rather the commission's "blessing" to act like a private business in terms of responding to the market. The risks of actual privatization, staff members felt, were too great. They predicted that a private operator would take the Milorganite name and apply it to sludges that were cheaper to produce. These other products would be "up-valued" while the Milorganite name would be eroded. When a contract terminated, the district would be "left with a bastardized product, no staff, and the continuing problem of how to dispose of the stuff" (Ward 1990). The marketing staff convinced the commission to keep the marketing operations public, and its predictions about private sludge companies proved prescient. Many of the private sludge marketers generated terrible publicity and ran into enormous opposition in disposing of their sludge. A number of firms were tied to organized crime, and others were prosecuted for bribing public officials, while their sludge was found to contain high concentrations of heavy metals and other pollutants (Stauber and Rampton 1995, 99–122).

The alternative to privatization was an increased commitment to developing and marketing new products. Milwaukee tried to license the Milorganite name to blends created by other fertilizer and chemical dealers. One such blend was a weed-and-feed product. (The district had seemingly forgotten its own history in inventing this product in the first place.) Milwaukee proposed to provide Milorganite to Parker Fertilizer, which would blend it with an herbicide and market it as "Milorganite Weed and Feed." A number of commissioners were concerned, however, that Milorganite's reputation as an "organic" or "natural" product would be damaged by "associating the Milorganite name . . . with '-cides' that may be environmentally unfriendly" (Approval to Enter into License Agreements 1990; Ralph Hollmon to Commissioners 1995; Alan Nees to Jim Hill 1995).

But marketing Milorganite as a natural product came under attack nonetheless in 1992 when the Federal Trade Commission began an investigation into Milorganite marketing. The Federal Trade Commission believed that Milorganite advertisements were deceptive by portraying Milorganite as "natural, organic, and/or safe." By 2000, the US Department of Agriculture explicitly banned all sewage sludge from use on crops labeled organic (New Rules on Organic Foods 2000). Another arena for marketing a new product was suggested by customers who noticed that Milorganite kept deer from browsing in their gardens. Deer, an increasingly damaging garden pest in suburban ecosystems, were repelled by the odor of Milorganite. EPA regulations on pesticides prevented the district from explicitly marketing Milorganite for this purpose without first registering the product under the Federal Insecticide, Fungicide, and Rodenticide Act, and Milwaukee undertook a decade-long process to register Milorganite as a deer repellant.

In addition to the problem of new product development, changes in Milwaukee's economy were causing a steep decline in Milorganite production. Milwaukee was caught between two opposing forces. It needed to innovate new products to replace the revenue lost to competition in its traditional golf course market (Behm 2006). At the same time, it needed to prevent steep declines in production to provide material to the markets it was busy creating. Milwaukee's investment in its Milorganite production capacity required a steady stream of sewage to produce the fertilizer. Inputs to the sewage treatment plant depended on developments in technology, economic conditions, and social movements, and in turn placed constraints on Milwaukee's ability to market its product.

Industrial waste played a contradictory role in Milorganite production. The district sought to prevent waste from the city's many machining and metal factories from entering the sewers and contaminating Milorganite with heavy metals. Yet other Milwaukee industries were vital to Milorganite production. As Milwaukee industry began a long economic decline in the 1970s, the closure of polluting factories in the metal industry helped lead to a cleaner sludge, with lower concentrations of heavy metals. But as the giant breweries Schlitz and then Pabst closed, followed by malting and yeast production plants, the sewage treatment plant was robbed of much of the raw material for its product (Rohde 2006). The city, famous for its beer, produced a great deal of nutritious waste that helped feed the microorganisms of the activated sludge process. When the breweries and yeast factories closed, all that material was lost to the plant, and Milorganite production declined. Milwaukee partly tried to make up for it by shipping sludge from its south side plant to Jones Island as well as contracting with neighboring communities to process their sludge (Sink 2008). The latest solution to the problem of decreasing supply involves trading on Milorganite's name and reputation. For over seventy years, Milwaukee's marketing staff has built up and protected the Milorganite name and brand. It may find that the name is more valuable than its own sludge. Milwaukee is seeking to license the Milorganite brand to other cities' sludge production.

Milwaukee's history of Milorganite production showed that sewage treatment plants could not be run at a profit. Even in its best years, when World War II had elevated the price of nitrogen by diverting chemical nitrogen into explosives production, the sales of Milorganite barely covered the cost of its own production, not to mention the cost of other parts of the plant (Utilization of Sewage Sludge as Fertilizer 1946, 98). Yet repeated economic analyses showed that once Milwaukee had sunk its initial capital investment into its fertilizer plant, continuing Milorganite production was a cheaper means of sludge disposal than using either a landfill or incineration. But even this depended on continuous innovation in product development and marketing. Milwaukee could not simply assume a market and sell to it. By pursuing an outlet for its sludge through the marketplace, Milwaukee was not only caught in the contradiction between profit and purification but the contradictions of capital as well. The pressures under capitalism for increasing competition, falling profits, and reduced market share demanded constant attention to the business aspects of sludge disposal. Only relentless innovation in product development and marketing allowed Milwaukee to dispose of its sludge.

But with the required emphasis on fertilizer production to meet costs, Milwaukee's entrance into the market distorted its goals of purifying its sewage. As Gail Radford (2003, 889) notes about the history of public authorities like the sewerage commission, entering the market often meant structural pressures to stress revenue at the expense of environmental concerns:

Agencies . . . are impelled by their very nature to behave much like private businesses. They have to pay bondholders, staff, and vendors from revenues gained mostly by charging for the goods and services they provide. Thus, they must make revenue growth and cost reduction the central criteria for deciding what to produce and how to do so. Only with difficulty can they give goals such as environmental protection . . . much weight.

In 1960, Raymond Leary, Milwaukee's chief engineer, stated, "We are not in business as a manufacturer, we are a sanitary facility to collect and treat sewage" (quoted in Milorganite Moves from Profit to Loss 1960). That this needed to be reiterated points to the fundamental contradictions of sludge and the distortions in public policy that were created when Hatton made sewage treatment subject to the relentless demands of the market.

Notes

1. See also Granville 1865.

2. For the history of these failed attempts to capitalize on sewage value, see Schneider 2011.

3. For a history of Milwaukee's water, see Foss-Mollan 2001.

4. For more details on the relation between Milwaukee politics and sewage, see Schneider 2011. See also Hamlin 1988.

5. Weed control plot, Milarsenite, Westmoor Country Club, October 23, 1941; chemical weed control plots, North Shore CC, Milwaukee, "Weed Control 1941."

6. Tentative letter to R. F. Kneeland, Food and Drug Administration, May 1, 1952.

7. The commission prevented the use of "Daorganite" (Miami) and "Muskegonite" (Muskegon), as well (Alan K. Nees to Bill Parker 1992).

8. Except where noted, the history of Rapidgro is constructed from various newspaper clippings contained in a scrapbook from the Grand Rapids treatment plant records. Most of the article clippings in the scrapbook, however, have neither a source nor date, although they appear to be chiefly from the 1930s. Series 16–20, Environmental Protection Department, Wastewater Treatment Plant, Miscellaneous Historical Records, Grand Rapids City Archives. .

9. Citizens for a Better Environment also protested Chicago's sludge, now called "Nu-earth," and forced the city to place warning signs where the city gave the sludge away to the public (Plan Alert Signs 1978).

10. A recent report suggests that brain trauma like that suffered by football players may mimic the symptoms of and be misdiagnosed as ALS. This report was based on studies of the neuropathology of yet another football player for the 49ers initially diagnosed with ALS (Schwarz 2010).

11. There remains great controversy surrounding the safety of sewage sludge, even that considered of exceptional quality. For a critique of sludge safety, see Stauber and Rampton 1995, 99–122.

References

Advertisement. 1931. *National Greenkeeper* 5 (4): 28.

Advertisement. 1932a. *National Greenkeeper* 6 (5): 15.

Advertisement. 1932b. *National Greenkeeper* 6 (9): 4.

Advertisement. 1932c. *National Greenkeeper* 6 (12): inside back cover.

Alan K. Nees to Bill Parker. 1992. February 20. Box 9000, Milwaukee Metropolitan Sewerage District.

Alan Nees to Jim Hill. 1995. November 27. Box 9000, Milwaukee Metropolitan Sewerage District.

Alexander M. McIver and Son to V. H. Kadish. 1926. December 9. Noer/Milorganite® MMSD Collection, Turfgrass Information Center, Michigan State University Library.

Annual Budget Report to the Board of Trustees. 1940. *Making Recommendations for 1940*. Chicago: Sanitary District of Chicago.

Approval to Enter into License Agreements with Selected Fertilizer Blenders. 1990. May 16.

Barron, Tom. 1992. Can Cities Turn Profits from Sludge? *Environment Today* 3 (4): 3–4.

Bazalgette, C. Norman. 1876–1877. The Sewage Question. *Minutes of Proceedings of the Institution of Civil Engineers* 48:105–251.

Behm, Don. 2006. Milorganite as Garden Deer Repellent? It Could Happen: Seeking Official Designation. *Milwaukee Journal Sentinel*, November 19.

Benefits Seen in Milorganite. 1952. Unidentified newspaper clipping, March 31. Milwaukee Sewerage Commission, 1933–1960. Xeroxed clippings, Milwaukee Public Library.

Brower, James. n.d. *Report: Resume of Twenty-Five Years of Plant Operation in Connection with Drying Sewage Sludge.* Files of Tom Brennan, Milwaukee Metropolitan Sewerage District.

B12 Vitamin Production from Activiated [sic] Sludge Under Way at Milwaukee Sewage Plant. 1956. *Water and Sewage Works* 103 (10): 477.

Cadmium Content Warning. 1980–1989. Box 7011. Milwaukee Metropolitan Sewerage District.

Caster, Arthur D. 1955. Cities Report Varying Success in Drying Sludge for Sale. *Wastes Engineering* 26 (11): 610–611.

Charles G. Wilson to William Katz. 1977. July 25. Box 3673. MMSD.

Collings, Gilbeart H. 1955. *Commercial Fertilizers: Their Sources and Use.* 5th ed. New York: McGraw-Hill Book Company.

Commercial Possibilities with Activated-Sewage Sludge. 1916. *Engineering Record* 74:428.

Copeland, William R., and John Arthur Wilson. n.d. Final Report of the Investigations and Experiments of Sewage Disposal Processes Made at Milwaukee, Wisconsin. Box 9171, Records Office, Milwaukee Metropolitan Sewerage District.

Crawford, C. W. to O. J. Noer. 1951. Personal correspondence, June 16. Box 923, Milwaukee Metropolitan Sewerage District.

Daniels, P. N. 1934. Increasing the Salability of Sludge. *Water Works and Sewerage* 81:277–279.

Eleventh Annual Report of the Sewerage Commission of the City of Milwaukee for the Year 1924. 1925. January 1.

Falvey, Pamela J., and Geoffrey Hurtado. n.d. Report on the Milwaukee Metropolitan Sewerage District Solids Utilization Department. Box 5708, Milwaukee Metropolitan Sewerage District.

Flynn, K. C. 1982. Sludge Marketing: The Quiet Revolution. *Journal of the Water Pollution Control Federation* 54:1267–1269.

Foss-Mollan, Kate. 2001. *Hard Water: Politics and Water Supply in Milwaukee, 1870–1995.* West Lafayette, IN: Purdue University Press.

Fortieth Annual Report of the Sewerage Commission of the City of Milwaukee for the Year 1953. 1954.

Forty-Fifth Annual Report of the Sewerage Commission of the City of Milwaukee for the Year 1958. 1959.

Forty-First Annual Report of the Sewerage Commission of the City of Milwaukee for the Year 1954. 1955.

Forty-Second Annual Report of the Sewerage Commission of the City of Milwaukee for the Year 1955. 1956.

Forty-Third Annual Report of the Sewerage Commission of the City of Milwaukee for the Year 1956. 1957.

Fowler to Bartow. 1915. December 14. Bartow Papers, University Archives, University of Illinois at Urbana-Champaign.

Granville, A. B. 1865. *The Great London Question of the Day; or, Can Thames Sewage Be Converted into Gold?* London: Edward Stanford.

Hamlin, Christopher. 1988. Muddling in Bumbledom: On the Enormity of Large Sanitary Improvements in Four British Towns, 1855–1885. *Victorian Studies* 32:55–83.

How Fertilizers Help to Establish Dense, Luxuriant Turf. 1927. April 18. Noer/Milorganite® MMSD Collection, Turfgrass Information Center, Michigan State University Library.

How to Use MILARSENITE for Clover and Weed Control on Golf Greens. n.d. Milarsenite leaflet no. 1, Weed Control Data + Correspondence. Noer/Milorganite® MMSD Collection, Turfgrass Information Center, Michigan State University Library.

Market Probe, Inc. n.d. Milorganite Market Positioning Study, External Interviews.

Marketing Study. 1994–1995. Box 9000, Milwaukee Metropolitan Sewerage District.

Milorganite Advisory Council Meeting. 1972. Toronto, Canada, November 19. Box 3673, Milwaukee Metropolitan Sewerage District.

Milorganite Advisory Council Meeting. 1976. Buena Vista, Florida, October 17. Box 3673, Milwaukee Metropolitan Sewerage District.

Milorganite Gardening: A Health Hazard. 1978. Citizens for a Better Environment Comment, March. Box 14128, Milwaukee Metropolitan Sewerage District.

Milorganite Market Positioning Study: Summary of Findings and Recommendations. 1994–1995. Marketing Study. Box 9000, Milwaukee Metropolitan Sewerage District.

Milorganite Moves from Profit to Loss. 1960. *Milwaukee Journal*, November 13. Milwaukee Sewerage Commission, 1933–1960. Xeroxed clippings, Milwaukee Public Library.

National Gardening Association. 1991. *National Gardening Survey, 1990–1991.* Burlington, VT: National Gardening Association.

National Gardening Association. 1996. *National Gardening Survey, 1995–1996.* Burlington, VT: National Gardening Association.

New Rules on Organic Foods. 2000. *New York Times*, March 9.

Nineteenth Annual Report of the Sewerage Commission of the City of Milwaukee for the Year 1932. 1933. January 2, 163–179.

Nitrogen Recovery from Sewage Sludge Reaches Commercially Practicable Stage. 1916. *Engineering Record* 74 (15): 444–445.

Noer, O. J. 1925. The Fertilizer Value of Activated Sludge with Particular Reference to Golf Courses. Progress Report for Seasons 1923 and 1924. Noer/Milorganite® MMSD Collection, Turfgrass Information Center, Michigan State University Library.

Noer, O. J. 1927. Annual Report of Milwaukee Sewerage Commission Fellowship, for Year Ending February 1, 1926. Noer/Milorganite® MMSD Collection, Turfgrass Information Center, Michigan State University Library.

Olis Proposes Package Sale of Fertilizer. 1954. *Chicago Daily Tribune*, April 23.

Plan Alert Signs at Nu-Earth Sites. 1978. *Chicago Tribune*, May 18.

Radford, Gail. 2003. From Municipal Socialism to Public Authorities: Institutional Factors in the Shaping of American Public Enterprise. *Journal of American History* 90:863–890.

Ralph Hollmon to Commissioners. 1995. November 6. Box 9000. MMSD.

Rawn, A. M. 1941. Concerning Those Plus Fertility Values of Sewage Sludges. *Water Works and Sewerage* 88:186–188.

Report 44–2 to the Sewerage Commission of the City of Milwaukee for the Period March 1 to May 1, 1994. 1944. Box 923, Milwaukee Metropolitan Sewerage District.

Report on the Attitude of Three Large Companies in the Feed Industry toward the Acceptability of Milorganite and Its Extracts as Vitamin B_{12} Supplements in Animal Feeds. 1952. December 17. Box 923, Milwaukee Metropolitan Sewerage District.

Report to the Board of Trustees. 1949. *Support of Recommendations for 1949 Budget*. Chicago: Sanitary District of Chicago.

Research: A National Resource II. 1941. *Industrial Research: December 1940*. Washington, DC: National Resources Planning Board.

Resolution Authorizing Approval to Modify the "Notice" on the Milorganite® 40 lb. Retail Bag. 1993. 93–165–9(02), September 27, Item 10, Box 9000. Milwaukee Metropolitan Sewerage District.

Robert J. Borchardt to David G. Berger. 1977. September 2. "Tech Bulletin 88," Noer/Milorganite® MMSD Collection, Turfgrass Information Center, Michigan State University Library.

Rockwell, Fulton. 1993. New York City Moves Ahead with Biosolids Management. *BioCycle* 34 (10): 55–60.

Rohde, Marie. 2006. A Greener Future? Fertilizer Slowdown Has MMSD Pondering Uses for Sludge. *Milwaukee Journal Sentinel*, September 14.

Royal Commission on Metropolitan Sewage Discharge. 1884. *Second and Final Report of the Commissioners*. C. 4253.

Sanitary District Will Demolish Fertilizer Manufacturing Plant. 1968. *Chicago Tribune*, December 1.

Sanitary Unit Can't Sell All Its Fertilizer. 1957. *Chicago Daily Tribune*, August 10.

Schneider, Daniel. 2011. *Hybrid Nature: Sewage Treatment and the Contradictions of the Industrial Ecosystem*. Cambridge, MA: MIT Press.

Schwarz, Alan. 2010. Study Says Brain Trauma Can Mimic ALS. *New York Times*, August 17.

Sink, Lisa. 2008. Brookfield Offers to Share Sewage Sludge. *Milwaukee Journal Sentinel*, August 18.

S. Shore Sewage Unit Urged Here. 1956. Unidentified newspaper clipping, March 21. Milwaukee Sewerage Commission, 1933–1960. Xeroxed clippings, Milwaukee Public Library.

Stauber, John, and Sheldon Rampton. 1995. *Toxic Sludge Is Good for You!* Monroe, ME: Common Courage Press.

Swanson to Hawkins. 1989. Milorganite Retail Bag: NOTICE. September 20. Box 7011. MMSD.

Tentative Program of Field Work for 1924 in Connection with the Commercial Development of Milwaukee's Activated Sludge. 1924. January 10. Noer/Milorganite® MMSD Collection, Turfgrass Information Center, Michigan State University Library.

Twelfth Annual Report of the Sewerage Commission of the City of Milwaukee for the Year 1925. 1926. January 1.

Utilization of Municipal Wastewater Sludge. 1971. Manual of Practice, no. 2. Water Pollution Control Federation.

Utilization of Sewage Sludge as Fertilizer. 1937. *Sewage Works Journal* 9:861–912.

Utilization of Sewage Sludge as Fertilizer. 1946. Manual of Practice, no. 2. Federation of Sewage Works Associations.

Vinton, W. Bacon to Robert J. Borchardt. 1977. July 6. "Tech Bulletin 88," Noer/Milorganite® MMSD Collection, Turfgrass Information Center, Michigan State University Library.

Vitamin B_{12} Test Project Is Contracted. 1953. Unidentified newspaper clipping, June 14. Milwaukee Sewerage Commission, 1933–1960. Xeroxed clippings, Milwaukee Public Library.

Ward, Terry. 1990. Inputs from Marketing Staff for Review. November 27. Box 7011, Milwaukee Metropolitan Sewerage District.

Ward, Terry. 1991. Marketing Task Force, February 25. Box 9000, Milwaukee Metropolitan Sewerage District.

Zahn, Michael. 1982a. Maryland Milorganite Ban Fought. *Milwaukee Journal*, August 15.

Zahn, Michael. 1982b. Safer Milorganite to Be Sold in Snazzier Bags. *Milwaukee Journal*, June 3.

8

The Rising Tide against Plastic Waste: Unpacking Industry Attempts to Influence the Debate

Jennifer Clapp

The recently discovered plastic "blob," a continent-size patch of garbage composed mainly of plastic waste floating in the Pacific Ocean, is a wake-up call for our modern consumptive society. It highlights the fact that much of the world's discarded plastic is not actually reused or recycled. Rather, it ends up as waste: disposed of in landfills, swirling in the ocean and other waterways, caught in trees, and littering roadsides. Plastics do not break down readily in the environment. The persistent presence of plastic waste in our surroundings, from local communities to the global commons, is a stark reminder of the growing ecological impact of our consumption and waste.

As environmental awareness has grown, people around the world have increasingly questioned the ubiquitous use of plastic, particularly as a packaging material to be readily discarded without much thought as to its final resting place. Initiatives have emerged in multiple locations to tax, ban, or otherwise regulate the use of certain types of plastic packaging, most prominently plastic bags and bottles. These regulatory changes regarding plastic in different communities, ranging from municipal- to state- to national-level measures, signal that small steps can add up to a larger shift in attitudes and practices concerning a specific type of waste. The change in norms has been both rapid and globally significant. But as regulatory initiatives spring up worldwide in response to the shifting public sentiment, representatives of the plastics industry have resisted these measures.

In this chapter, I examine the shift in norms and attitudes toward plastics along with the role of the plastics industry in resisting regulation that has emerged in response to new norms. I argue that as an antiplastic norm and associated regulatory measures have arisen, industry actors have actively resisted them, using several strategies. First, they have taken up a discursive campaign in the public domain in an attempt to assert that plastics are the most environmentally sound packaging choice. Second, they have acted to both lobby and litigate within communities seeking plastics regulation in an effort to impose a form of "regulatory chill." The idea of a regulatory chill was originally identified in reference to countries failing to raise environmental standards for fear that investors would leave for other jurisdictions (see Neumayer 2001). The failure of communities to enact legislation on local environmental problems for fear of litigation represents a similar dynamic of industry influence on public environmental policy.

Undergirding the plastics industry's specific strategies on plastic bags and bottles has been its firm belief in the rights and responsibilities of individual consumers. The right to choose a product and responsibility to properly dispose of its packaging, in industry's view, belongs with the individual consumer. Adherence to this principle is in line with the "individualization of responsibility"—a particular strand of environmentalism predominant in mainstream North America that sees both the problem and solution to environmental issues as being in the hands of individual consumers, rather than the producer or society as a whole (Maniates 2001, 33). This approach leaves little room for collective attempts to address environmental problems through, for example, regulation on the sale and distribution of plastics. Although industry has actively pursued its strategies to resist the regulation of plastic packaging over the past decade, the outcomes have thus far been highly uneven.

Communities Take Action on Plastic Waste

Within a short period of time, it has become increasingly socially unacceptable to use disposable plastic shopping bags, and in a growing number of places around the world it is now illegal to distribute them for free. Legislation is also restricting the purchase of plastic water bottles in many communities. A number of problems have been associated with plastic packaging waste, ranging from environmental degradation to concerns related to the potential health impacts of exposure to plastics. These issues have contributed to an emergent shift in people's views regarding the usefulness of plastic packaging. An interesting element of this shift in norms is that it emerged first in the Global South, challenging the widely held perception that developed nations are generally environmental leaders. As early as 1992, for example, citizens protested for tighter regulation of plastic bags in Dhaka, Bangladesh. Today, a growing number of communities in the Global North are seeing similar kinds of citizen engagement on this issue. This change in attitudes has emerged not as the product of a globally networked campaign against plastics but rather, interestingly, as a series of localized movements in different communities worldwide responding to specific local concerns (Clapp and Swanston 2009). Though local, these shifts have global significance.

A key problem with plastic packaging, which has contributed to many of the locally specific concerns about the waste it generates, is its permanence. Plastics can take hundreds, if not thousands, of years to degrade. Moreover, items such as plastic bags do not actually biodegrade. Instead, they "photodegrade"—meaning that they break down into smaller and smaller pieces, but do not disappear entirely. In the words of Charles Moore (2006), an oceanographer who studies plastic waste in the seas, "Plastic, like diamonds, are forever!" The plastic that went into producing every plastic bag ever handed out and every water bottle made is still out there, somewhere. It could be in the ocean, a tree, or a stream, along a roadside, at the bottom of a landfill, or in a bird's stomach, or it may have been recycled into something else, like a plastic lumber deck, or even more plastic bags and bottles.

Although many types of plastics can be recycled, the problem is that for the most part, they are not. Total recycling rates reflect both a "capture" of the material to be recycled and the demand for that material to be remade into other products. A mere 5 percent of plastic bags in North America, for example, are collected for recycling, and less than 25 percent of plastic water bottles are recycled.[1] The market for postconsumer plastic bags is weak. Even in nations with a developed recycling infrastructure, plastic bags are rarely included in major recycling programs, and if recycling of this kind of plastic is attempted at all, it is usually in the form of drop-off centers at retail locations. The main issue is contamination, which makes plastic bags much less attractive to recyclers than other types of plastic. The market for plastic water bottles is much stronger because of the type of plastic and lower contamination levels. But even in the case of water bottles, the vast majority of them are not collected for recycling—signaling that there are still major issues with the capture of this kind of plastic waste. The growing presence of discarded plastic in the environment, particularly after the rapid growth in plastics use in packaging over the past thirty years, indicates that recycling alone has not solved this problem. This reality has prompted communities around the world to consider management strategies other than recycling, such as action to restrict the use of plastic packaging in the first place.

One of the earliest manifestations of shifting public norms around plastic packaging was a growing distaste with plastic shopping bags

(Clapp and Swanston 2009). Single-use plastic shopping bags only really started to be handed out to shoppers in industrialized countries in the 1980s, and the developing world followed suit in the 1990s. These bags were seen to be convenient and inexpensive, and it quickly became standard for retailers and vendors to offer up a plastic bag to carry one's purchases home.[2] The lower cost combined with the ease of transportation and storage meant that plastic displaced the paper bags that retailers used to offer their customers. Yet the ease of offering plastic bags that followed meant that they quickly became ubiquitous. Shoppers who took them soon became inundated with plastic bags. Because manufacturers began to make them thinner and thinner in order to save costs, plastic bags soon were viewed as entirely "disposable," and made their way into the waste stream. In other words, convenience had become compatible with disposability, and the use of plastic bags became widespread with little consideration of the environmental and social consequences. Numerous reports note that the average amount of time a plastic bag is in a person's possession after receiving it from a retailer is approximately twelve to fifteen minutes—meaning that bags go from being a carrier of goods to useless waste with remarkable speed.[3]

Although presented as a convenience, plastic bags are a nuisance when they are carelessly discarded into the environment. A major concern is that they pose a hazard to wildlife. Marine animals such as sea turtles and fish as well as seabirds often mistake pieces of plastic bags for food, which can block their digestive systems and cause death. These animals can also get tangled up in them. Additionally, plastic bags have been implicated in clogging sewer drains in a number of countries around the world, contributing to flood risk. In warm climates, as has been documented in several African countries, they can act as a breeding ground for malaria. Plastic shopping bags cause litter problems because they can blow around easily, making their way into streams, trees, and bushes (Environment Australia 2002), and are particularly prevalent on the shores of waterways.

Rising discontent with the problems associated with plastic bags spurred legislative changes in many communities around the world. South Asia was the first area to legislate a restriction on the use of plastic shopping bags. In the 1990s, the national government in Bangladesh and several states in India began to adopt outright bans on the distribution and use of plastic bags. Concerns about flooding were a key motivator for action in Bangladesh in the late 1990s, after plastic bags were found to have blocked sewer drains. The 1998 floods inundated two-thirds of the country and were responsible for hundreds of deaths (see Reazuddin 2006). In India, concerns about sacred cows, which had been dying from ingesting plastic bags that blocked their intestines, was a major motivating factor for restrictive legislation in a number of states and cities across the country (Krulwich and Goldstein 2000). Other parts of the developing world started to adopt similar legislation soon after, including a number of countries in eastern and southern Africa and East Asia.[4] The South African government, for example, introduced the idea of a plastic bag ban in 2001, and in 2002 the Plastic Bag Agreement was reached, imposing a ban on ultrathin bags and a levy on thicker, more durable bags. In Taiwan, a phased-in ban of plastic bags was implemented 2002–2003 in response to concerns about toxic emissions from incineration.

Jurisdictions in the industrialized world also started worrying about plastic bags in the early to mid-2000s. In 2002, Ireland adopted a steep tax on plastic shopping bags, which some claim reduced plastic bag use by 90 to 95 percent in a short period of time (see Convery, McDonnell, and Ferreira 2007). Since 2000 several cities in Australia imposed bans, and the national government adopted a voluntary program to reduce plastic bag use in 2008 (Environment Australia 2002; Groves 2008). By 2007, in North America and the United Kingdom, proposals emerged for action against plastic bags. San Francisco was the first US city to ban bags, followed by Los Angeles County, the city of Los Angeles, and a number of smaller towns in California (Clapp and Swanston 2009). In the United Kingdom, over eighty towns and villages instituted a voluntary retailer ban on handing out plastic bags, and the UK prime minister at the time, Gordon Brown, went so far as to say that if retailers did not cut their use of the bags, the national government would consider legislation (Brown Threatens 2008). In 2008, China adopted a national-level ban on plastic bags, and although street vendors often ignore the ban, a 2009 industry report suggested that plastic bag usage dropped by two-thirds in the ban's first year (China Bans 2008; Watts 2009). In Canada, several small towns have instituted bans, and Toronto adopted a surcharge starting in 2009 on each plastic bag handed out by retailers (City of Toronto 2009).

The distaste for plastic packaging has manifested not just in an abrupt turn away from the use of plastic shopping bags. Plastic water bottles are also increasingly being shunned in a growing number of communities around the world. Most bottled beverages prior to the 1990s were packaged in glass, but there was a major shift to plastic bottles in the 1990s—a trend led by soft drink manufacturers Coke and Pepsi (see Clarke 2007). Water for sale in plastic bottles has taken off with remarkable growth in recent decades. Globally, some 180 billion liters of bottled waters were sold in 2007, compared to only 78 billion in 1997 (Milmo 2007). In the United States alone, 31 billion single-use plastic water bottles were sold in 2007, up from just 3 billion in 1997 (New York State Department of Environmental Conservation n.d.).

The waste implications of the ballooning sales of bottled water have been staggering. Because less than one-quarter of water bottles are actually collected for recycling in the United States, they have come to litter public spaces such as parks and beaches, and also make their way into waterways (Container Recycling Institute 2009). Some people attribute this low recycling rate to the fact that, unlike plastic bottles for other beverages, there has, until recently, rarely been a deposit charged on plastic water bottles. Recycling rates for plastic water bottles are likely to be much lower in other countries where recycling and collection facilities are not as advanced as in North America.

Plastic water bottles have additional baggage attached to them when compared with plastic shopping bags, thereby contributing to a shift in norms around their use. Not only does the bottle in which water is sold have serious implications for the waste stream; the packaging of water for sale in the first place raises concerns for many about the sale of what is considered a basic need (Clarke 2007). Attitudes toward plastic have also influenced human health concerns, particularly with respect to potential chemical leaching from the plastic into the water (Biello 2008). Energy considerations have also been prominent in shaping norms. Bottled water requires an enormous amount of energy for bottling and transportation from its source, as it frequently crosses borders and travels long distances, with one-quarter of all bottled water crossing an international border (Larsen 2007).

The environmental problems associated with plastic water bottles are many, especially considering that municipalities worldwide strive to provide clean and safe drinking water at a fraction of the cost of bottled water. In response, many municipalities have banned the sale of bottled water in municipal buildings and schools. As far back as 1987, the city of Los Angeles adopted regulations that restricted the purchase of bottled water with city funds. Other cities around the world have adopted similar rules in recent years. In 2007, for example, San Francisco began to prohibit the purchase of bottled water by city departments and agencies (City and County of San Francisco 2007). Vancouver and Charlottetown in Canada, Ann Arbor and Santa Barbara in the United States, and Florence, Italy, among others, have enacted similar legislation, while scores of other towns and cities are considering adopting restrictions on the purchase of bottled water by the local government as well as its sale in city-owned buildings (Larsen 2007). In June 2008, the US conference of mayors passed a resolution to encourage a phaseout of government use of bottled water where feasible,[5] and in March 2009, the Federation of Canadian Municipalities asked Canadian cities and towns to phase out the purchase and sale of bottled water on city property (CBC 2008).

Some communities have gone further than just banning the sale of bottled water on municipal property. The City of Chicago (2008) instituted a five-cent tax on plastic water bottles in 2008, and New York State imposed a deposit on water bottle sales in 2009 (New York State 2009). The state of Michigan is currently considering a ten-cent tax on bottled water to fund education scholarships (Mui 2009). In 2009, the Australian town of Bundanoon became the first community in the world to ban the sale of bottled water outright—even going so far as to remove it from store shelves. The town's residents were upset not just about the waste implications but also by the fact that a bottled water company had announced plans to tap water in the community, truck it to Sydney for bottling, and then ship it back to sell in Bundanoon (Foley 2009; BBC 2009). In January 2011, the town of Concord, Massachusetts, voted to ban all bottled water sales, (Abel and Woods 2010).

Reinforcing the emergence of new norms and policies regarding plastic bags and water bottles, the discovery in the late 1990s of what's been dubbed the Great Pacific Garbage Patches has both fascinated and frightened people about the long-term impact of plastic waste (Weisman 2007). Ninety percent of the waste in the patches, each double the size of Texas, stretching across twenty-seven hundred kilometers of the Pacific Ocean, is comprised of plastic wastes. The plastic collects in gyres—areas in the ocean where currents meet—in which waters swirl; some also have described the gyre as "a giant toilet bowl with no drain" (Fraser 2008). Most of the plastic waste in these gyres has become translucent from exposure to the sun. Much of it floats just beneath the water's surface, although most of it has probably sunk much deeper, making its measurement particularly difficult. The United Nations Environment Programme, however, has estimated that there are approximately thirteen thousand pieces of plastic waste found in every square kilometer of sea, with higher concentrations in the garbage patches (Newey 2009). Seventy percent of that plastic eventually sinks below the surface and down to the seabed (Weiss 2006).[6]

Some of the top items found in marine litter surveys are cigarette filters, plastic bags, beverage bottles, plastic cutlery, food containers, and straws. Around 10 percent of plastic waste in the oceans consists of what are called "nurdles"—tiny plastic beads, which are the stock material for the plastics industry. The vast majority of the waste in the ocean gyres, some 80 percent, was discarded on land and then made its way to sea. This would explain why most of that waste is plastic; it is light and travels easily in water. Another troubling fact is the leaching of toxic chemicals from the plastic into the water (Barry 2009). Further, when sea animals mistake the floating bits of plastic for food, they become contaminated when they eat it, and in turn, contaminate the food chain, including humans who eat the fish. Large marine animals, such as turtles, albatross, and whales, can also become entangled in the waste floating in the garbage patches, or perish due to blockages in their stomachs from bits of plastic that become stuck in their digestive tract (UNEP 2009).

In the background to the growing awareness that plastic waste poses in the oceans and within communities, there is also a small but growing awareness of the "ecological shadows" associated with the global trade in recyclable plastic wastes. Ecological shadows are the aggregate environmental impact of one country's activity in relation to domestic consumption and the avoidance of its own environmental problems that is felt in other parts of the world (Dauvergne 2008). Transporting plastic waste around the world for recycling creates plenty of ecological shadows. The export of recyclable plastics from municipal recycling programs in industrialized countries to developing countries such as China and India is an example of the way in which ecological impacts generated by the activity of one region are felt in another region. Over fifty countries export plastic waste to China, for instance (Sun 2008). Due to the revelation in many communities in recent years that the plastic waste placed in household recycling collection boxes for curbside recycling programs has actually been shipped to China, a number of municipalities have banned the export of plastic waste, as occurred recently in several Canadian cities (Outhit 2007). The main concerns have been with the conditions under which these wastes are recycled as well as the energy consumed in transporting the wastes long distances, especially when local recycling facilities are struggling to stay in business.

A key reason that plastic wastes have been transported globally for recycling is the strong demand from rapidly industrializing countries such as India and China for secondhand plastic material. For these countries, it is much less expensive to use recycled rather than virgin plastic in their manufacturing; it is also easy to tap into the active global market in plastic waste. In the United States, for example, more used plastic bottles are exported than recycled domestically (NAPCOR 2008). Exports of postconsumer PET plastic bottles have increased dramatically since the 1990s—by a factor of ten between 1998 and 2008 (ibid.). Over 57 percent of the PET plastic bottles collected in the United States for recycling are exported, with most of those making their way to China, and nearly a quarter of HDPE plastic bottles are also exported (American Chemistry Council 2008). One-third of the plastic bags collected in North America are exported too—and again, most make their way to locations in Asia. Most of the plastic bags collected in municipal curbside recycling programs are exported because these bags are more likely to be contaminated, making them more difficult to recycle, but also less expensive to import for recyclers in countries such as China (Stewardship Ontario 2007).

The growth in the various legislative initiatives around the world to address plastic waste over the past decade demonstrates that communities are becoming increasingly aware of the problems associated with the excessive use of plastic packaging. There is rising awareness that plastic waste disposal and recycling—what we normally think of as a locally specific issue—has not just local or regional impact but also global environmental implications. Plastic waste affects the global commons such as the oceans and contributes ecological shadows that transcend international borders. Because waste is typically legislated locally, communities have been able to adopt regulatory measures to suit their own specific needs and concerns, and have done so relatively quickly. Web sites promoted by some local community and environmental groups have also emerged to educate citizens on the impacts of plastic waste, promote alternatives such as reusable bags and water bottles, and share ideas with other communities on how to tackle these issues locally.[7] Collectively, these developments signal a major shift in norms surrounding plastic waste.

Industry Fights Back

Although communities worldwide have taken legislative action at the local level to tackle some of the problems linked to plastic waste, the road toward regulating these wastes has been contested. Industry actors—the plastics industry as well as business interests in the bottled water industry—have been prominent in campaigns to resist regulation on plastic shopping bags and water bottles. They have actively sought to shape the discourse around the issue, prominently promoting the alternative norm of "recycling" as the most environmentally sound response to the problem. They have also actively lobbied and litigated to reverse or prevent legislation that aims to curb use of the plastic packaging in question.

Underlying industry's approach to resisting plastics regulation has been its strong adherence to the principles of individual consumers' right to choose what products they consume and individual responsibility for what ends up in the waste stream. Instead of placing responsibility for the environmentally sound management of plastic packaging on producers, industry instead has attempted to put it squarely on consumers. In objecting to regulation, industry has also tried to appeal to the idea of consumer freedom. If consumers are simply educated to "do the right thing," there should be no need for regulation, and consumers' right to choose their packaging preferences is preserved. This approach is consistent with the individualization of responsibility within certain strands of environmentalism. As Michael Maniates (2001, 33) insightfully notes regarding the trend toward individualization within North American environmentalism, "Education is a critical ingredient in this view—smart consumers will make choices, it's thought, with the larger public good in mind."

In most countries where the call for plastic bag restrictions has become prominent, the plastics industry has launched a full-out campaign, forming a number of associations with the specific agenda to protect and defend the plastic bag. Each of these industry associations has posted informational Web sites and distributed public relations materials to make the case that plastic is a superior packaging material when compared to its alternatives. Such claims rest centrally on the unproven assumption that banning plastic bags will result in their substitution with paper bags, and that plastic bags represent less embedded energy, and by extension carbon emissions, than their paper counterparts.

The main industry groups that have cropped up to defend plastic bags include the Progressive Bag Affiliates (PBA), formerly a subgroup of the American Chemistry Council, the largest chemicals industry lobby group in the United States. The PBA moved its affiliation to the Plastics Industry Trade Association (SPI) in 2012. PBA members include plastic bag manufacturers as well as other large firms such as Dow and Exxon Mobil.[8] The PBA sponsors a Web site that claims to offer facts about plastic bags. The Film and Bag Federation, a member of the SPI, hosts a Web site that seeks to correct myths about plastic bags.[9] The Save the Plastic Bag coalition is yet another group that sponsors a Web site defending plastic bags. Lawyer Stephen Joseph, who *Time* magazine recently dubbed the "patron saint of plastic bags," fronts for the group, made up primarily of plastic bag manufacturers (Luscombe 2008). In the United Kingdom, the

Carrier Bag Consortium hosts a Web site to defend the plastic bag and argue against legislation such as bag taxes.[10]

Proplastic Web sites present a particular discourse around plastic bags. They seek to convey competing norms that industry has judged will appeal to consumers as much, if not more, than that of turning away from plastic bags altogether. These contentions are grounded in the individualized concept of environmentalism discussed above, and judiciously avoid the very concerns—permanence, unsightly litter, and habitat degradation—that put plastics on the agenda in the first place. Perhaps the most prominent argument made by industry, and a direct attempt to challenge the new norm that "plastic is bad for the environment," is that plastic bags are more environmentally sound than other bag alternatives. In order to make this case, industry first asserts that paper bags are the main alternative to plastic bags. It then draws on life cycle analysis assessments to demonstrate that plastic bags are environmentally superior to paper bags because they are more energy efficient to produce, using 70 percent less energy in their manufacture than paper bags.[11] They are also lighter and easier to transport than paper, which offers even further savings on greenhouse gases because fewer trucks are needed to carry the same number of bags. In addition, the manufacture of plastic bags uses far less water than is the case with paper bags.[12] There is little discussion of the comparative energy-use impacts of reusable bags, and no evidence that paper bags are the main alternative to plastic ones.

The industry also maintains that plastic bags have other environmental benefits. Plastic bags supposedly take up less space in landfills because they are lightweight and compact. Furthermore, they purportedly emit fewer greenhouse gases in landfills because they do not decompose, and thus they do not release carbon like paper bags do when they decompose. According to industry information posted on Web sites, plastic is also more efficient to recycle than paper because it uses 91 percent less energy in the recycling process. Plastic bags, the industry points out, can be recycled into a number of useful new products, including plastic lumber for decks and fences, other building and construction products, and more plastic bags.[13] The poor recycling rates of plastic packaging are strangely absent from industry's information on the preferred strategy of plastic recycling.

Plastic bags, according to the plastics industry, are more sanitary than reusable ones. In early 2009, the Environment and Plastics Industry Council, a subdivision of the Canadian Plastics Industry Association, sponsored a study on reusable shopping bags that showed high bacteria counts in reusable bags, using the information to argue that plastic bags are safer (EPIC 2009). This contradicts another industry claim, however, that plastic bags are not a problem because the can be reused, and industry even advocates the reuse of plastic bags to carry groceries.[14] A further reason that plastic bags are more sanitary, in the plastics industry's view, is that paper bags attract cockroaches. According to the Save the Plastic Bag Coalition, Ireland experienced a rise in cockroach infestations following the imposition of its plastic bag tax.[15] Yet this claim contradicts other ones made by industry, noted below, that the plastic bag tax in Ireland did not reduce plastic bag use.

At the same time that they praise the virtues of plastic bags, the plastics industry representatives vehemently deny environmentalists' charges about the dangers of plastic bags. They question allegations that plastic bags cause harm to wildlife, holding that environmental groups have no solid proof that marine animals are harmed by plastic bags, and that pieces of plastic easily pass through animals' digestive systems if ingested.[16] Industry also denies that the bags form a major component of the waste stream, pointing out that they make up only 8 percent of coastal litter.[17] These claims ignore recent scientific evidence from both independent researchers and international organizations that identify plastic marine debris as a major global environmental issue (see Moore 2008; UNEP 2009).

Finally, the plastics industry has also devoted a significant portion of its Web presence to criticizing policies that seek to curtail the use of plastic bags. Industry claims that taxes and bans are misguided because they do not accurately reflect the environmental situation, and that such policies do not work to cut plastic bag use.[18] The Irish plastic bag tax is frequently cited as a failed policy, because plastic bag use, according to the plastics industry, rose by 10 percent after the tax was imposed, and because the tax placed hardships on both retailers and consumers.[19] This is in contrast to other reports that plastic bag use dropped dramatically as a result of the tax and that support for the tax is widespread (Rosenthal 2008). The bag ban in San Francisco is portrayed by industry as a failure, since it resulted in more paper bags being handed out, which results in higher greenhouse gas emissions, and more litter.[20]

Instead of regulatory approaches such as bans or taxes, industry has argued that the only way to reduce litter is to educate consumers to change their individual behavior so that they recycle more. One industry Web site puts it bluntly: "Educating and encouraging consumers to make environmentally-conscious decisions about plastic bags is a practical alternative to imposing taxes during a recession, or banning plastic bags altogether."[21] In other words, rather than accepting the responsibility, industry is advancing the idea that individuals should take primary responsibility for the problem.

In making these various assertions in its attempt to shape the discourse around plastic bags, industry has skirted the permanence issue, which is a primary reason communities sought to ban these bags in the first place. Industry has denied that plastic bags are a significant part of the waste stream and a major component of litter. It has also contended that when plastic bags do end up in a landfill, the fact that they do not decompose is a good thing, because then they do not emit greenhouse gases. Industry has ignored outright the issue of toxic chemicals leaching into water and the food chain. It has deflected attention from these other concerns by focusing on climate change as opposed to the other environmental and social problems associated with plastic bags. In short, industry has tried to make people think it is only a choice between paper and plastic, and that energy and resource use in the manufacture is all that matters, rather than the hazards linked to plastic's makeup and its longevity in the environment.

The second part of the plastic industry's strategy has been to influence legislation more directly. It has lobbied municipal governments in an attempt to head off any consideration of bag bans in North American towns. In Canada, for instance, after the small town of Leaf Rapids banned plastic bags in 2007, the Canadian Plastics Industry Association (2007) undertook an email campaign—directed toward all municipal councillors across the country—vowing to publicly oppose anyone who takes a stand against plastic bags. In 2009, the city of Philadelphia was also subjected to heavy lobbying by the Progressive Bag Affiliates when it sought to propose a ban on plastic bags. Retailers also joined in the lobbying, and under this pressure the city eventually dropped its pursuit of a bag ban (Thompson n.d.). Industry groups in numerous California towns have undertaken similar lobbying campaigns.

As part of its campaign to influence legislation, the plastics industry has gone further than lobbying. It has sponsored litigation and threats of litigation against a growing number of towns enacting as well as considering enacting legislation on plastic bags. This strategy has been a deliberate attempt to not only repeal existing bans on plastic bags but also create a regulatory chill to prevent other communities from considering bans. Industry's main approach has been to file lawsuits against municipalities on the grounds that they did not conduct their own environmental impact assessment prior to deciding on or proposing legislation to restrict plastic bag use. A number of municipalities have backed off their legislative plans for fear of the high costs associated with a lawsuit against them.

The Save the Plastic Bag coalition prominently features a litigation section on its Web site, with a timeline showing the various lawsuits it has launched and their progress.[22] It has launched formal suits as well as filed legal objections with at least eight municipalities in California that have enacted or proposed plastic bag bans. The first of these was a lawsuit against the small town of Manhattan Beach. The suit states that the city failed to conduct an environmental impact assessment as required by the California Environmental Quality Act for all actions that would have a substantial environmental impact. It argued that had the city of Manhattan Beach conducted this assessment, it would have determined that paper bag use would go up, and that paper bags are worse for the environment than plastic, based on the industry's own assessment as posted on its Web site.[23]

Manhattan Beach was surprised when the Los Angeles Superior Court ruled in February 2009 that the city should have conducted an environmental impact assessment prior to implementing its legislation and ordered it to conduct a more in-depth study of the paper versus plastic issue. The city's attorney in the case noted that the city still did not agree that it should have conducted a full assessment prior to implementing the legislation on plastic bags, because these assessments are only required if substantial environmental impact is expected from an action—something that the city did not think was likely (Dryden 2009). While the city's attorney did not deny that paper might have a higher carbon footprint, he did not agree that paper bag use necessarily increases as a result of plastic bag restrictions. He focused his arguments around the issue of permanence: the fact that paper biodegrades, while plastic does not. The court decision in this case may impact the outcomes of the other lawsuits launched by the Save the Plastic Bag Coalition, and may well have a chilling effect on other cities also considering advancing legislation.

In recent years, the plastics industry has gone further in its attempt to prevent communities in California from imposing bans on plastic bags by proposing statewide legislation that would outlaw such bans. In early 2009 the PBA, for example, proposed bill AB 1141, which while imposing a small fee on producers for each bag provided to retailers (for a total fund that would be capped at twenty-five million dollars) to be channeled toward litter abatement, would also prohibit other governmental bodies in the state from banning plastic bags, including nullifying existing bans that have been challenged in court (American Chemistry Council 2009a). This bill ultimately did not make it out of the California Assembly Natural Resources Committee (Los Angeles Solid Waste Management 2009, 9).

Industry Defends Water in Plastic Bottles

The defense of bottled water has been important to the beverage industry due to the phenomenal growth in bottled water sales over the previous decade. Local action against bottled water has thus been seen as a threat— all the more so because it has begun to take a bite out of the market. In the first half of 2009, bottled water sales fell for the first time in five years (Mui 2009). The beverage industry, rather than the plastics industry, has taken the lead in defending the sale of water in plastic bottles. Similar to the strategy pursued by the plastics industry regarding plastic bags, the bottled water industry has sought to shape discourse by promoting the environmental benefits of bottled water. It has also attempted to lobby and litigate against bottled water bans. Given that the debate over bottled water is wider than just the packaging issues, the industry has made arguments on a number of fronts, including the health implications of bottled versus tap water.[24] The focus here, however, is on those industry claims that specifically touch on the plastic waste aspects of bottled water.

The main actors defending bottled water are the US-based International Bottled Water Association (IWBA) and its local chapter in Canada, the Canadian Bottled Water Association. Other associations, such as Refreshments Canada, have been active in Canada. Several coalition organizations have also been active in seeking to influence the debate, including the Web site Enjoybottledwater.org, which is sponsored by the US-based Competitive Enterprise Institute. Additionally, the IWBA formed a "coalition" to defend bottled water with the Web site Bottlewatermatters.com, although the Web site does not list any actual coalition members. In late 2009, Bottled Water Matters began to upload videos to YouTube that defended bottled water in an effort to appeal to younger people, who may be influenced by the anti-bottled-water campaigns that have cropped up on environmental groups' and social networking Web sites. Other pro-bottled-water blogs and Web sites have also appeared on the Internet, including I Love Bottled Water, which features a "Bottled Water Girl of the Month," presumably designed to attract visitors.[25]

Industry has focused on a number of arguments to defend bottled water in popular discourse. To start with, it claims that banning sales of bottled water will not reduce the number of plastics bottles that end up in landfills. Instead of seeking a drinking fountain to satisfy their thirst, industry maintains that people will simply purchase other drinks, also packaged in plastic bottles, so that the amount of plastic waste resulting from beverages will remain the same. Nestlé Waters Canada, for example, claims that it has conducted its own survey of Canadian citizens, revealing that 60 percent of the respondents reported that they would buy other drinks if bottled water was not available. When the city of London, Ontario, voted to ban bottled water in municipal buildings, Refreshments Canada, an industry lobby representing the snack food industry in Canada, stated that this move was a "step backwards for recycling," and "will cost taxpayers more and do less for the environment" (CBC 2008). Because waste issues related to packaging affect all commercial beverages sold in plastic bottles, the industry has complained that it is unfair to target bottled water versus other drinks.

The beverage industry stresses that plastic water bottles, made of PET plastic, are fully recyclable with superior recycling efficiency when compared with glass or other packaging materials.[26] Furthermore, it notes that plastic water bottles are among the most recycled packaging in the United States (even though less than 25 percent of those bottles are recycled).[27] Organizations such as the IBWA and Refreshments Canada advertise that they support recycling infrastructure in order to educate consumers to recycle. The IBWA Web site clearly targets individual consumers' responsibility to recycle in its promotion of initiatives to "improve curbside recycling efforts and increase recycling at parks, sporting venues, other on-the-go locations, and in the home and office."[28] Taking this approach firmly steers the discourse away from industry's role in manufacturing demand for bottled water in the first place, and avoids any discussion of reuse or recycling systems that demand greater involvement from industry such as a deposit return system.

Industry also claims that even those plastic water bottles that are not recycled do not make up a major component of litter. Representatives repeatedly allege that less than one-third of 1 percent of all waste in the United States is made up of water bottles, although it is unclear where exactly this information originates.[29] They further assert that landfills are now bigger and better, such that consumers should not worry about the lack of landfill space for discarded water bottles.[30]

In terms of a carbon footprint, the bottled water industry is quick to point out that it is using lighter plastics to increase transportation efficiency. Nestlé Waters, for example, highlights the fact that it has reduced the weight of its plastic water bottles by 24 percent between 2004 and 2009 and that recycling itself saves energy over the use of virgin packaging materials.[31] Indeed, bottled water advocates, like advocates of plastic bags, point to life-cycle assessment studies, contending that plastic is 47 percent more energy efficient than aluminum cans and 63 percent more energy efficient than glass.[32] Despite claims that plastic is superior environmentally, however, one advocacy Web site oddly appears to validate the desire to reduce plastic water bottles by maintaining that "even if we could eliminate bottled water altogether, global climate would remain unaffected," because to reduce warming would require major economic changes that are unlikely to come about anytime soon. It asserts that to ban bottled water under such circumstances would be futile, and only end up denying choice to consumers.[33]

Notwithstanding efforts to paint itself green in terms of bottled water's carbon footprint, it is difficult for proponents to demonstrate clearly that the product has lower carbon emissions than tap water—the main alternative. Tap water activists have argued that bottled water, even in lighter plastic bottles, uses over a hundred times the greenhouse gases of tap water to provide. Some bottled water advocates recognize that they are on thin ice with respect to this issue, and have begun to advertise that they purchase carbon offsets to make up for their carbon emissions. Fiji Water, for instance, pledged in 2007 to be "carbon negative" by reducing its carbon emissions, increasing its reliance on renewable energy sources, and offsetting its remaining carbon emissions by 120 percent (Mui 2009). In 2011 the company was sued over this claim, and the company removed the pledge from its website (Bloxham 2011).

Like the plastics industry with respect to bags, proponents of bottled water have dedicated a significant amount of their efforts to critiquing bans and taxes as unworkable solutions that deny consumers choice. A recent report sponsored by the Competitive Enterprise Institute, and posted on the Enjoy Bottled Water Web site, calls bottled water regulations an assault on consumer freedom that takes "the Nanny state to a whole new level" (Logomasini 2009).

As in the case of plastic bags, the bottled water industry has also sought to directly influence legislation through lobbying and litigation where bans on bottled water have been proposed. Municipalities that have debated bans and taxes on bottled water sales have had to face the full force of lobbyists. According to Tony Clarke (2008), a prominent Canadian anti-bottled-water activist, when Toronto debated a bottled water ban in municipal facilities in 2008, "a battery of lobbyists, corporate executives and industry associations" descended on city hall to try to influence the vote. Similar pressure was applied to city councillors in other Canadian municipalities.[34] New York State also met with intense lobbying as it considered its bottle bill, and Michigan also faced heavy lobbying from the bottled water industry when it proposed to institute a bottled water tax in 2010.

In some cases, industry has sought to influence city councillors by offering a carrot instead of wielding a stick. In Ontario, for example, Nestlé Waters said it would pay for new recycling programs targeting the collection of plastics in public spaces in those communities considering bottled water bans. But this offer was only available to those communities if they rescinded their plans to ban bottled water in municipal facilities.[35] This carrot, combined with a threat to remove the offer if bottled water bans went ahead, was not appreciated by city councillors in Waterloo, London, Toronto, and St. Catharines—all of which went ahead with plans to ban bottled water in municipal facilities.

In other instances, industry has taken a tougher stance. In cases where cities or states have imposed bottled water taxes or deposit laws on water bottles, bottled water industry associations have launched lawsuits. When Chicago implemented its bottled water tax in early 2008, the IBWA along with the American Beverage Association, Illinois Retail Merchants Association, and Illinois Food Retailers Association launched a lawsuit against the city for its imposition of the tax, which they claim will cost forty-five hundred jobs and drive consumers outside the city to do their shopping.[36] Their legal case against the tax is that it is a tax imposed on food, which is exempt from tax within the city, and thus they claim that Chicago had no jurisdiction to impose the tax.[37] At the time of this writing, there had not yet been a ruling on this case. Meanwhile, the IBWA also joined Nestlé Waters and Polar Corp to file a federal lawsuit against the New York bottle bill on a number of grounds, including that the bill does not require deposits on bottled water with sugar added (Truini 2009). The IBWA announced in 2010 that it was considering the option of a legal challenge against Concord, Massachusetts, in response to its vote to ban bottled water sales. The town reversed its decision in 2011.[38]

Which Norms Will Carry the Day?

Local, national, and regional governmental authorities worldwide have responded to growing anti-plastic-waste sentiments among their populations in recent years by enacting legislation to curb the use of plastic packaging. These regulatory efforts are often closely tied to local concerns over litter and waste as well as a growing worry about the ecological effect of plastics in the global commons and around the world. A new norm has begun to emerge that calls for a drastic reduction in the consumption of the most ubiquitous and least necessary forms of plastic packaging: plastic shopping bags and water bottles. Local authorities, including municipalities and other subnational governmental authorities, have frequently played a leading role in driving change and thus have been able to enact legislation that reinforces norms in a relatively short period of time. But bringing about such change in legislation at the local level, especially in North America, has met strong resistance from industry actors with vested interests in the continued use of plastic packaging.

Industry actors—the plastics industry in the case of plastic bags, and the bottled water industry in the case of plastic bottles—have tried to influence the debate over plastics. They have attempted to shape the broader discourse, on the one hand, and lobby and litigate to achieve more direct legislative influence, on the other hand. In their discursive strategies, industry actors have presented competing environmental norms to the antiplastics norm that has arisen globally of late. These competing norms include recycling and energy efficiency. By appealing to these alternative environmental norms, the industries in question are seeking to downplay the environmental hazards associated with plastic packaging waste, which stem largely from its permanence, and instead portray plastic packaging as environmentally sound based on life cycle assessment analyses that focus on energy efficiency when compared with specific alternative forms of packaging such as paper or glass.

Although these alternative norms are appealing in principle to most environmentalists on some level, it is questionable whether these arguments prevail over other concerns linked to plastic waste. Most communities that have targeted plastic bags and bottles have done so based on the permanence of plastics in the environment, which is precisely what makes plastic a hazard, and the fact that most plastic waste does not in fact end up in the recycling stream in actual practice. In short, industry's efforts to shape the environmental discourse have not won over the activist groups seeking to bring about change on these issues. Industry's resistance to the antiplastic norm has challenged environmental and community groups to clarify their positions as well as rationale for legislative action, which only strengthens their position. The broader public may find some attractive elements to industry's arguments, but it is as yet unclear whether this has led a majority of citizens to demand that their local governments provide them the right to use these plastic products. Many people have switched their practices away from plastic shopping bags and water bottles willingly, without need of legislation.

When the strategy of shifting discourse in favor of these alternative norms has not been sufficient to stop legislation from being adopted in local communities, industry has shown that it is quite willing to play tough to get what it wants. In both the bottles and bags cases, industry has not only offered carrots such as funding for litter abatement and recycling infrastructure in return for repealed legislation but also launched lawsuits that sought to roll back legislation that regulates both bottles and bags. Industry has been successful in its legal strategy in some cases, but not in others.

In its various responses to the plastic waste issue, industry has consistently portrayed itself as a champion of consumer rights and responsibilities. It has shown support for individuals' right to stop using plastic, but would not tolerate broader regulation that it saw as infringing on the "rights" of consumers to use plastic packaging if they wished. In resisting legislation that curtailed plastics use, industry also promoted the idea that responsibility for addressing the problem lies at the individual consumer level. If plastic bags and water bottles are not recycled, it is because people are not putting them in their recycling bins. This stance is evident in the plastics industry's recent response to the problem of plastics in the oceans. The American Chemistry Council spearheaded a public private partnership in the state of California that places five hundred recycling bins on Californian beaches. The program stresses on its media page that "plastics do not belong in our oceans and waterways—they belong in the recycling bin" (American Chemistry Council 2009b). The notion of this partnership is to ensure that consumers place their waste in recycling bins, in direct opposition to the idea of many communities to restrict the use of plastic packaging in the first place.

Industry's attempt to individualize responsibility for plastic waste by trying to influence the discourse along with the pursuit of legislative action on plastics has been actively resisted by a growing number of communities, which instead have pursued the legislation even in the face of industry's attempts to stop them. A key reason for the continued community push on plastics is one that industry has failed to acknowledge: the plastic packaging that communities are targeting—plastic shopping bags and water bottles—is in most cases completely unnecessary. Industry has expressed puzzlement over why these particular forms of plastic packaging have been singled out for legislation. But the answer is simple: water bottles and carrier bags were targeted first because they are items for which there are straightforward alternatives. As such, it is relatively easy to get people to change their behavior in a short amount of time. And the strategy is working. The shift to reusable packaging forms for water and shopping has been significant, even in cases where these items have not been restricted by legislation. In the United Kingdom, for example, even without widespread legislation on plastic bags, use fell by 26 percent between 2006 and 2008 (Gray 2009).

Some have argued that efforts to regulate plastics are miniscule in light of the huge environmental problems linked to today's growth-centered, highly unequal, overconsumptive society (Dauvergne 2010). But while broader structural change indeed could help to address the ecological impacts of consumption, including the plastic packaging that encases much of that consumption, the absence of political will for broader changes should not dissuade community efforts to reduce plastic waste. Though small in impact with respect to the broader problem of waste in our environment, the shift in norms in relation to plastic bags and water bottles is an important first step toward more expansive measures to deal with the growing menace of plastic packaging waste that threatens the planet's health. The fact that a growing number of communities have proposed collective responses to plastic packaging waste signals that larger transformations governing consumption's ecological impacts are possible even in the face of industry resistance.

Notes

I would like to thank Brittney Martin and Linda Swanston for superior research assistance. I would also like to thank the Social Sciences and Humanities Research Council of Canada for financial support for this research. All Web pages cited were accessed February 3, 2012.

1. American Chemistry Council 2008; US EPA 2006. See also Container Recycling Institute 2009.

2. On the rise of the norm of convenience in modern society, see Shove 2003.

3. See, for example, Ford 2008.

4. Clapp and Swanston 2009; Hasson, Leiman, and Visser 2007; McLaughlin 2004; UNEP 2005; BBC 2007.

5. See US Conference of Mayors Web site: http://usmayors.org/usmayornewspaper/documents/06_16_08/pg6_res_environment.asp.

6. See also UNEP 2001.

7. See, for example, Californians against Waste, http://www.cawrecycles.org/issues/plastic_campaign/plastic_bags; Algalita Marine Research Foundation, http://www.algalita.org; Polaris Institute, http://www.insidethebottle.org/.

8. For the Web sites sponsored by the American Chemistry Council, see http://plastics.americanchemistry.com/Market-Teams/Plastics-in-Packaging.

9. For the PBA Web site, see http://www.plasticbagfacts.com/. For the Plastics Industry Trade Association's Web site on plastic bags, see http://www.plasticsindustry.org/AboutPlastics/content.cfm?ItemNumber=712&.

10. For the Carrier Bag Consortium Web site, see http://www.carrierbagtax.com/.

11. For PBA news releases, see http://www.plasticbagfacts.com/.

12. See http://www.plasticbagfacts.com/Main-Menu/Fast-Facts.aspx.

13. Ibid.

14. See the Plastics Industry Trade Association Web site, http://www.plasticsindustry.org/IndustryGroups/content.cfm?ItemNumber=520.

15. See the Save the Plastic Bag Coalition Web site, http://www.savetheplasticbag.com/ReadContent598.aspx.

16. Ibid.

17. See the Bag the Ban Web site, http://www.bagtheban.com.

18. See Carrier Bag Consortium, http://www.carrierbagtax.com/downloads/rubbish.pdf.

19. See http://www.plasticbagfacts.com/Main-Menu/taxes-and-bans-dont-work.

20. Ibid.

21. See http://www.plasticbagfacts.com/Main-Menu/Reduce-Reuse-Recycle.

22. See http://www.savetheplasticbag.com.

23. For the text of the lawsuit, see http://www.savetheplasticbag.com//UploadedFiles/STPB%20LA%20County%20Complaint.pdf.

24. For these debates, see Clarke 2007.

25. See http://www.squidoo.com/private_label_bottled_water.

26. See, for example, the Refreshments Canada Web site, http://www.refreshments.ca/19-media-centre/41-letters-to-the-editor/76-march-31-saskatoon-star-phoenix-re-bottled-water.

27. See the IBWA Web site, http://www.bottledwater.org/content/recycling-and-environment.

28. Ibid.

29. Ibid.

30. See the Enjoy Bottled Water Web site: http://enjoybottledwater.org.

31. See http://www.nestle-waters.com/environment/reduce-co2-emissions/greenbottle-packaging.html.

32. Ibid.

33. Ibid.

34. See, for example, Weeks 2008.

35. Observation by the author at the Waterloo Regional Council meeting where such offers and threats were made.

36. Demirjian 2007. See also the IWBA Web site, http://www.bottledwater.org/content/lawsuit-filed-overturn-chicago-bottled-water-tax.

37. For the complaint, see http://www.bottledwater.org/public/downloads/2008-01-04_bw_complaint.pdf.

38. See the IBWA's press release, http://www.bottledwater.org/news/ibwa-applauds-citizens-concord-ma-defeating-proposed-ban-bottled-water.

References

Abel, David, and Jason Woods. 2010. Concord Fires First Shot in Water Battle. *Boston Globe*, May 1. http://www.boston.com/news/local/massachusetts/articles/2010/05/01/concord_fires_first_shot_in_water_battle/.

American Chemistry Council and Association of Postconsumer Plastic Recyclers. 2008. *2008 United States National Post-Consumer Plastics Bottle Recycling Report*. http://legis.wisconsin.gov/lc/committees/study/2010/SUP/files/08_USpostconsumer_plasticsbottlerecyclingreport.pdf

American Chemistry Council. 2009aPalo Alto, CA, Rejects Partnership to Increase Recycling, Outlaws Plastic Bags. |Press Release. http://www.americanchemistry.com/Media/PressReleasesTranscripts/RelatedPDF/Palo-Alto-CA-Rejects-Partnership-to-Increase-Recycling-Outlaws-Plastic-Bags-PDF.pdf

American Chemistry Council. 2009b. Plastics Makers Call for Enhanced Education and Waste Infrastructure to Address Marine Debris. Press Release. http://www.americanchemistry.com/Media/PressReleasesTranscripts/ACC-news-releases/Plastics-Makers-Call-for-Enhanced-Education-and-Waste-Infrastructure-to-Address-Marine-Debris.html.

Barry, Carolyn. 2009. Plastic Breaks Down in Ocean, After All—and Fast. *National Geographic News*, August 20. http://news.nationalgeographic.com/news/2009/08/090820-plastic-decomposes-oceans-seas.html.

BBC. 2007. East Africa Ban on Plastic Bags. *BBC News*, June 14. http://news.bbc.co.uk/2/hi/africa/6754127.stm.

BBC. 2009. Australia Town Bans Bottled Water. *BBC News*, July 8. http://news.bbc.co.uk/2/hi/asia-pacific/8141569.stm.

Biello, David. 2008. Plastic (Not) Fantastic: Food Containers Leach a Potentially Harmful Chemical. *Scientific American*, February, 19.

Bloxham, Andy. 2011Fiji Water Accused of Environmnetally Misleading Claims. *The Telegraph*, June 20. http://www.telegraph.co.uk/earth/earthnews/8585182/Fiji-Water-accused-of-environmentally-misleading-claims.html.

Brown Threatens Supermarkets over Plastic Bag Reduction. 2008. *Times* (London), February 28.

Canadian Plastics Industry Association. 2007. In Defence of Plastic Shopping Bags. PowerPoint presentation. On file with author.

CBC. 2008. London, Ont., to Ban Bottled Water on City Premises despite Beverage Industry Protests. *CBC News,* August 19. http://www.cbc.ca/news/story/2008/08/19/london-bottledwater.html.

China Bans Free Plastic Shopping Bags. 2008. *International Herald Tribune,* January 8. http://www.iht.com/articles/2008/01/09/asia/plastic.php.

City and County of San Francisco. 2007. *Permanent Phaseout of Bottled Water Purchases: Facts for City and County of San Francisco Employees.* San Francisco: City and Country of San Francisco. http://www.wellnesswater.ca/Articles/Bottled-WaterPhaseOut.pdf.

City of Chicago. 2008. *Chicago Bottled Water Tax Guide.* Chicago: City of Chicago Revenue Department. http://www.cityofchicago.org/content/dam/city/depts/rev/supp_info/TaxSupportingInformation/BottledWaterTaxGuide.pdf.

City of Toronto. 2009. *Plastic Shopping Bags Cost a Minimum of 5 Cents as of June 1, 2009.* http://www.toronto.ca/garbage/packaging_reduction/5centbag_by-law.htm.

Clapp, Jennifer, and Linda Swanston. 2009. Doing Away with Plastic Shopping Bags: Explaining International Patterns of Norm Adoption and Policy Diffusion. *Environmental Politics* 18 (3): 315–332.

Clarke, Tony. 2007. *Inside the Bottle: An Expose of the Bottled Water Industry.* Ottawa: Polaris Institute.

Clarke, Tony. 2008. Toronto Stood Up to the Bottled Water Industry. *Toronto Star,* December 11, 11. http://www.thestar.com/comment/article/551909.

Container Recycling Institute. 2009. *Wasting and Recycling Trends: Conclusions from CRI's 2008 Beverage Market Data Analysis.* Culver City, CA: Container Recycling Institute. http://www.container-recycling.org/assets/pdfs/reports/2008-BMDA-conclusions.pdf.

Convery, Frank, Simon McDonnell, and Susana Ferreira. 2007. The Most Popular Tax in Europe? Lessons from the Irish Plastic Bags Levy. *Environmental and Resource Economics* 38:1–11.

Dauvergne, Peter. 2008. *The Shadows of Consumption: Consequences for the Global Environment.* Cambridge, MA: MIT Press.

Dauvergne, Peter. 2010. The Problem of Consumption. *Global Environmental Politics* 10:1–10.

Demirjian, Karoun. 2007. Retailers to Fight Bottled Water Tax; Food, Beverage Alliance to Sue Chicago over Levy. *Chicago Tribune,* December 27. http://articles.chicagotribune.com/2007-12-27/news/0712260740_1_bottled-water-international-bottled-water-association-american-beverage-association.

Dryden, Carley. 2009. Judge Axes Plastic Bag Ban, City Ponders Their Next Move. *Easy Reader,* February 26. http://archive.easyreadernews.com/story.php?StoryID=20034722&IssuePath=.

Environment and Plastics Industry Council (EPIC). 2009. *Grocery Carry Bag Sanitation: A Microbiological Study of Reusable Bags and "First or Single-Use" Plastic Bags.* http://www.carrierbagtax.com/downloads/Microbiological_Study_of_Reusable_Grocery_Bags.pdf.

Environment Australia. 2002. *Plastic Shopping Bags: Analysis of Levies and Environmental Impacts.* Final report. Victoria: Environment Australia. http://www.tud.ttu.ee/material/piirimae/eco-design/Plastic%20bag/analysis-final.pdf.

Foley, Meraiah. 2009. Bundanoon Journal; Ban on Bottled Water, Apparently a First, Puts Small Town on Big Stage. *New York Times,* July 17. http://query.nytimes.com/gst/fullpage.html?res=9A0CEFDC133FF934A25754C0A96F9C8B63.

Ford, Matt. 2008. Fighting the Tide of Plastic Bags in a World Awash with Waste. CNN, July 19. http://edition.cnn.com/2008/WORLD/asiapcf/07/13/eco.plasticbagwaste/index.html.

Fraser, Stephen. 2008. What a Dump. *Current Science* 94 (2): 6–7.

Gray, Louise. 2009. Plastic Bag Use Falls by 26 Percent in Two Years. *Telegraph,* February 26. http://www.telegraph.co.uk/earth/earthnews/4807924/Plastic-bag-use-falls-by-26-per-cent-in-two-years.html.

Groves, S. 2008. Bagging the Bag a State of Mind. *Bay Post,* November 19. http://www.batemansbaypost.com.au/news/local/news/general/bagging-the-bag-a-state-of-mind/1364356.aspx.

Hasson, Reviva, Anthony Leiman, and Martine Visser. 2007. The Economics of Plastic Bag Legislation in South Africa. *South African Journal of Economics* 75 (1): 66–83.

Krulwich, Robert, and Jessica Goldstein. 2000. India Cow Killer Bagged, but Deaths Continue. National Public Radio. http://www.npr.org/templates/story/story.php?storyId=91310904.

Larsen, Janet. 2007. Bottled Water Boycotts: Back-to-the-Tap Movement Gains Momentum. Earth Policy Institute. http://www.earth-policy.org/plan_b_updates/2007/update68.

Logomasini, Angela. 2009. *Bottled Water and the Overflowing Nanny State: How Misinformation Erodes Consumer Freedom.* Washington, DC: Competitive Enterprise Institute. http://cei.org/cei_files/fm/active/0/Angela%20Logomasini%20-%20Bottled%20Water%20and%20the%20Overflowing%20Nanny%20State.pdf.

Los Angeles Solid Waste Management Committee and Integrated Waste Management Task Force. 2009. *Inside Solid Waste.* Los Angeles: Los Angeles Solid Waste Management Committee. http://ladpw.org/EPD/tf/isw/isw_2009_07.pdf.

Luscombe, Belinda. 2008. The Patron Saint of Plastic Bags. *TIME Magazine,* July 27. http://www.time.com/time/nation/article/0,8599,1827021,00.html.

Maniates, Michael. 2001. Individualization: Plant a Tree, Buy a Bike, Save the World? *Global Environmental Politics* 1 (3): 31–52.

McLaughlin, Kathleen. 2004. Wrap That in Plastic? Not in Taiwan, Unless You Pay. *Christian Science Monitor,* June 15. http://www.csmonitor.com/2004/0615/p07s02-woap.html.

Milmo, Cahal. 2007. Backlash against Plastic Bottles as Sales of Water Hit 1 Billion Litres. *Independent,* September 18.

Moore, Charles J. 2006. Out in the Pacific Plastic Is Getting Drastic. http://marine-litter.gpa.unep.org/documents/World's_largest_landfill.pdf.

Moore, Charles J. 2008. Synthetic Polymers in the Marine Environment: A Rapidly Increasing, Long-term Threat. *Environmental Research* 108:131–139.

Mui, Ylan Q. 2009. Bottled Water Boom Appears Tapped Out: Environmental Concerns, Recession Put Crimp in Sales. *Washington Post*, August 13.

National Association for PET Container Resources (NAPCOR). 2008. *Report on Postconsumer PET Container Recycling Activity*. Sonoma, CA: National Association for PET Container Resources. http://www.napcor.com/pdf/2008_Report.pdf.

Neumayer, Eric. 2001. Do Countries Fail to Raise Environmental Standards? An Evaluation of Policy Options Addressing "Regulatory Chill." *International Journal of Sustainable Development* 4 (3): 231–244.

New York State Department of Environmental Conservation. 2009. *New York's Bottle Bill*. http://www.dec.ny.gov/chemical/8500.html.

New York State Department of Environmental Conservation. n.d. *Too Many Bottles: It's a Waste*. http://www.peanc.org/sites/default/files/root/waterbottles.pdf.

Newey, Guy. 2009. Voyage to the Center of the "Plastic Vortex." *Agence France Presse*, May 24. http://www.google.com/hostednews/afp/article/ALeqM5hrjgT-1KiDZJmEkNDedH-0ZXmVb5g.

Outhit, Jeff. 2007. Region's Recycling in China Prompts Probe. *Waterloo Region Record*, August 25.

Reazuddin, M. D2006. *Banning Polyethylene Shopping Bags: A Step Forward to Promoting Environmentally Sustainable Development in Bangladesh*. Dhaka: Bangladesh Center for Advanced Studies.

Rosenthal, Elisabeth. 2008. Motivated by a Tax, Irish Spurn Plastic Bags. *New York Times*, February 2. http://www.nytimes.com/2008/02/02/world/europe/02bags.html?_r=3&em&ex=1202101200&en=b81a97080dcd4b15&ei=5087.

Shove, Elizabeth. 2003. *Comfort, Cleanliness, and Convenience: The Social Organization of Normality*. New York: Berg Publishers.

Stewardship Ontario. 2007. *Ontario Plastics Market Development Strategy*. Ontario: Stewardship Ontario. http://www.stewardshipontario.ca/bluebox/pdf/funding/plastics/Plastics_Market_Dev_Report.pdf.

Sun, Nina Ying. 2008. Worldwide Recycled Plastic Trade Hits the Rocks. *Plastics and Rubber Weekly*, November 11. http://www.prw.com/subscriber/newscat2.html?cat=1&channel=320&id=1227001849.

Thompson, Isaiah. n.d. Meet Big Bag: The Plastic Bag Industry Has Its Own Lobby, And in Philly, It's Been Damn Effective. 2009. *Philadelphia City Paper*, December 22. http://archives.citypaper.net/articles/2009/12/24/plastic-bag-lobby-philadelphia.

Truini, Joe. 2009. Nestlé Waters Files Lawsuit Challenging N.Y.'s Bottle Bill. *Plastics News*, May 22. http://www.wasterecyclingnews.com/scrap_recycling/scrap-report.html?id=1243611765.

United Nations Environment Programme (UNEP). 2001. *Marine Litter: Trash That Kills.* http://www.unep.org/regionalseas/marinelitter/publications/docs/trash _that_kills.pdf.

United Nations Environment Programme (UNEP). 2005. *Selection, Design, and Implementation of Economic Instruments in the Solid Waste Management Sector in Kenya: The Case of Plastic Bags.* http://www.unep.ch/etb/publications/ EconInst/Kenya.pdf.

United Nations Environment Programme (UNEP). 2009. *Marine Litter: A Global Challenge.* http://www.unep.org/regionalseas/marinelitter/publications/docs/Marine_Litter_A_Global_Challenge.pdf.

United States Environmental Protection Agency (US EPA). 2006. *Municipal Solid Waste in the United States: Facts and Figures.* http://www.epa.gov/osw/nonhaz/ municipal/pubs/msw06.pdf.

Watts, Jonathan. 2009. China Plastic Bag Ban Has Saved 1.6m Tonnes of Oil. *Guardian*, May 22. http://www.guardian.co.uk/environment/2009/may/22/china-plastic-bags-ban-success.

Weeks, Carly, 2008. First It Was Plastic Bags; Now Water Bottles Are the New Faux Pas. *Globe and Mail*, August 18.

Weisman, Alan. 2007. *The World without Us.* New York: Thomas Dunne Books.

Weiss, Kenneth. 2006. Altered Oceans: Plague of Plastic Chokes the Seas. *Los Angeles Times.*

Time Out of Mind: The Animation of Obsolescence in *The Brave Little Toaster*

Marisol Cortez

Since the 1970s, garbage—the visible remains of individual acts of consumption—has been one of the most recognizable public faces of environmental destruction. The target of public service announcements and government programs, from the US Forest Service's classic call to "Give a Hoot, Don't Pollute!" to the Texas Department of Transportation's ongoing "Don't Mess with Texas" campaign, postconsumer waste exists within the dominant environmental imaginary as an object of contempt and loathing—something that from an early age, we learn to regard as morally and politically abhorrent.

One of the well-documented downsides of this affective relation to waste is the way it tends to individualize environmental problems. The hypervisibility of the soda cans carelessly tossed on to the highway shoulder, for example, obscures the larger problem of industrialized food production and distribution systems, with their reliance on both packaging and long-distance transport. In some cases, propagation of this confusion between waste as hated object and the systemic nature of environmental problems has been intentional, as in the famous case of the 1970s' "Crying Indian" ad. Produced by Keep America Beautiful, a front group for the bottling industry, this ad presented an ostensible American Indian (in actuality played by Italian American actor Iron Eyes Cody) who weeps at the site of a US landscape despoiled by garbage and smog. With its tagline "People start pollution. People can stop it," the ad directed a burgeoning environmental consciousness away from corporate activity and toward the actions of individuals instead (Strand 2008). Dominant social relations to waste as garbage are thus marked by a central contradiction or problematic, in which waste either disappears from consciousness altogether, or where it appears, evokes such intense negativity that it also disappears from consideration the historical and social forces that have shaped how we relate to it, and hence the question of whether other, less ecologically destructive relations might exist.

In *The Ethics of Waste: How We Relate to Rubbish*, cultural critic Gay Hawkins explores similar problematics, arguing that environmental moralizing about waste frequently reproduces the very relations to nature it tries to change. As she observes, cultural contempt for waste objects is motivated by broader assumptions about the disenchantment of nature within urban industrial society as well as our alienation as humans from this nature. But this assumption in fact reinforces this alienation along with the radical separation of nature and culture it presumes. Writing about the dancing plastic bag scene in the film *American Beauty*, Hawkins (2006, 31) is disturbed that "a major environmental problem [is] rendered sensuous and enchanted," yet at the same time finds this aestheticization captivating, contrasting it with antilitter campaigns that present plastic bags as contaminating "the purity and otherness of the environment." "How is it," she asks, "that a scene from a hit movie has more emotional and political impact on me than the EPA's waste education campaign? Why is it that the EPA commercial leaves me feeling guilty and patronized, irritated by its explicit pedagogical intent? Could it possibly be more 'environmentally friendly' to feel sympathy and ethical concern for rubbish rather than shame, disgust, and anxiety about it?" What happens, she wonders, when *waste* rather than nature or even people become "the motivation for new actions" (ibid., 133)?

This chapter considers another key example from popular culture where the literal enchantment or bringing to life of waste suggests an alternate affective basis for environmental ethics and praxis than an individualizing waste hatred: the 1987 film *The Brave Little Toaster*. A low-budget animation based on a novella by sci-fi writer Thomas M. Disch, *Toaster* is a sort of machine-age *Incredible Journey*, in which five anthropomorphized appliances—a toaster, radio, bedside lamp, electric blanket, and vacuum cleaner—travel great distances to reunite with the human "master" who has seemingly abandoned them.

Yet overlaid onto this otherwise-sentimental narrative is something darker and stranger, something much more in keeping with the film's sci-fi origins: a dystopian vision of consumerism illuminated not by its human or even ecological casualties but more unpredictably by its spent commodities. For in the course of their journey, the appliances discover that they have become obsolete—"out of time" in the sense of being at the end of their socially useful lives. As with the dancing plastic bag in *American Beauty*, this portrayal of waste is deeply sympathetic; we are moved by and enchanted with the image of waste—the supposed evidence of nature's instrumentalization and disenchantment within industrial capitalism—as itself enchanted, a literal animation entirely hidden from view of the human characters within the film.

This trope of imagining the secret lives of inanimate objects is not unique to *Toaster*. Almost by definition as a genre, animation (which in the United States is largely a children's genre) is concerned with the relationship between human and nonhuman worlds, and in particular between humans and inanimate objects. From the long-running *Thomas and Friends* to more recently popular shows like *Bob the Builder* and *Handy Manny*, children's culture is rife with narratives about the literal animation of machine technology in particular—a fantasy of industrial objects that speak, sing, emote, and serve their human masters with a childlike eagerness to please.

Toaster is distinctive, however, in spearheading a more recent trend within broader generic emphases on animating the inanimate. For within the sunny tendency that includes *Bob*, *Thomas*, and *Manny* is a darker thread: stories that narrate the deaths as well as lives of objects, bringing to light the plight of obsolescence for the anthropomorphic objects of children's culture, and by extension, larger ecological problematics of waste and disposal within capitalist modernity. To some extent, obsolescence themes have always been part of the broader set of children's narratives that animate the ordinary objects of industrial culture. Stories like Margery Williams's *The Velveteen Rabbit* (1922) derive much of their poignancy from the figure of a once-loved object being cast off as garbage, its status as mere thing revealed by and revealing of the ruthlessness of an emergent industrial logic of accumulation and disposal that was changing the way that people not only related to "nature" but to the material world of everyday objects as well.

In a more recent era of rapid environmental change and mounting anxieties about its human causes, we see a further proliferation of narratives about animated objects that are also largely organized around themes of obsolescence and disposal. Pixar's breakout film *Toy Story* (1995), for example, seems almost to recap its supersession of Disney's traditional 2-D animation by computer graphics technology in its presentation of the conflict between toy cowboy Woody and high-tech action figure Buzz Lightyear. The studio's later, massively popular film *WALL-E* (2008) furthers these themes of obsolescence in explicitly environmentalist terms, telling the story of a trashed Earth where human and plant life alike have become impossible due to the manipulations of the megacorporation Buy n Large. This chapter concentrates on *Toaster*, then, not only for reasons of personal attachment but also because the film is arguably an important reflection of larger historical tensions between consumer culture and children's culture, and a crucial forerunner to a more recent, explicitly environmental turn within animation specifically, as the dominant form of children's culture in the United States.

Still, my focus here is less these historical tensions or children's culture broadly, and more the way popular representations of waste objects evoke particular affective responses, and in doing so, construct particular kinds of subjectivity in relation to an economic logic of disposal or displacement. It would be easy, given this emphasis, to view *Toaster*'s enchantment of consumer discards—its vision of appliances that sing, dance, move, and talk, thereby captivating film audiences—as the very literalization of commodity fetishism. According to the original conceptualization of the term by Karl Marx (1978, 325), commodity fetishism refers to the condition within commodity cultures like capitalism in which relationships between things seem to acquire a fantastic secret life of their own, eclipsing the social relations between those who labored to produce them. It is a "perceptual disorder," as political theorist Jane Bennett (2001, 13) describes it, or a condition of enchantment in which the fantastic quality of commodities obscures the suffering engendered in their histories of production (and futures of disposal).

What I would propose, however, is that *Toaster* dramatizes rather than simply literalizes commodity fetishism. That is, it performs how commodity fetishism functions as a specifically *temporal* displacement, in which we come not to be able to "see" either the life history of individual objects from the cradle to the grave, or the combined effects of this history on human bodies and nonhuman nature. That objects are out of time in this way makes them appear entirely independent of natural or social context—as autonomous beings without past or future, and whose appearance of individuality animated films then imagine as personhood. Yet in performing commodity fetishism as temporal displacement, the film at the same time destabilizes and problematizes it, and even provides its corrective through our affective investment in the story of objects whose "lives" are threatened by processes of commodification, consumption, and disposal. For unlike later films like *WALL-E*, it is not just the plight of a polluted Earth or humanity threatened by that pollution but instead waste itself that becomes the object of our empathy and concern.

It is this difference that I want to explore here; like Hawkins (2006, 22), I am interested in what is at stake for an environmental ethic of waste in the invitation to "change our affective relations with [and] delight in something [we] have been trained to hate." What kinds of subjectivity do

Toaster's waste sympathies engender, and what kinds of ethical practice or agency are made possible (or inhibited) by these ways of being with and knowing waste? As I argue, in bringing to life the idea that waste has a "secret life" and suggesting the significance of discerning this life, *Toaster* offers an alternate materialism more akin to that of German philosopher Walter Benjamin or the practice of life cycle assessment than a more traditionally Marxist critique of consumerism and mass culture. While seemingly disparate bodies of knowledge, Benjamin and life cycle analysis nonetheless both represent critical traditions of revisioning individual objects in terms of the histories (and futures) that their commodification contains yet hides from view. *Toaster* channels both of these traditions in illuminating the temporality of objects ordinarily obscured by the global and systemic nature of capitalist social relations. As such, the film's waste sympathy proposes an environmental ethic centered not on "getting rid" of waste as a hated object but rather on cultivating a *mode of perception* that recognizes and affirms all the ways in which we have a continuing relationality to what we consume as well as discard, even as powerful economic and social forces work to obscure these lines of connection.

Out of Time, Out of Mind

For those who haven't seen it, *The Brave Little Toaster* is a traditionally drawn, 2-D animated film whose tone and feel are similar to feature-length Disney films of the same era: colorful, its characters cartoonish and stylized rather than naturalistic, its humor light, and its story line interspersed with musical numbers. Now marketed as a Disney "classic," its relation to Walt Disney Animation Studios is actually quite complicated. Disney originally purchased the film rights to Disch's sci-fi novella shortly after its 1980 publication, with animator John Lasseter intending to develop the film as a feature-length, computer-generated images project. Disney's management rejected this idea for cost reasons, firing Lasseter (who went on to develop Pixar Studios into the animation giant it is today) and transferring production to Hyperion Pictures, an independent animation studio founded by two Disney expats (Paik and Iwerks 2007, 39–40). At the same time, Disney retained the video and television rights for the film—an acquisition that the company exploited when it premiered *Toaster* on its new cable network, even as the independent distributor Skouras Pictures slated the film for theatrical release. *Toaster* thus had a limited theatrical run, appearing primarily on the Disney Channel before going to home video (Schweiger 2004, 18).

This brief production history accurately maps how my early personal encounters with the film intersect with its circulation as a cultural object. A staple of my late 1980s childhood, I first encountered *Toaster* in its repeat screenings on cable television, and its sympathetic treatment of obsolete appliances made a deep impression on me, one renewed on viewing the film again as an adult. As a child, I could hardly have articulated the strange fascination I had (and continue to have) for this aspect of *Toaster*; particularly striking to me was the scene of the film's climax— a city dump—in which stacks of junked cars waiting their turn for the trash compactor reminisce in song about lives now pronounced worthless. Something about this scene was so eerie, so disturbing, and yet so *moving* that it captivated me in a way that even now I don't know quite how to express, as did the dump scene's culmination, in which Toaster throws himself into the jaws of the trash compactor in order to save his master (hence the reference to bravery in the film's title).

I begin my reading of *The Brave Little Toaster* with these personal memories, because the sort of sensuous attachment I have to the film as an object—my enchantment as a viewer with its image of enchanted waste objects—is precisely what the film as a narrative explores in its focus on the relationship between master and household appliance, child and beloved object. This relationship of attachment is also what I argue *Toaster* proposes as a model for an alternate ethics of waste and consumption.

In *The Enchantment of Everyday Life*, Bennett (2001, 8–9) has outlined this alternative ethics in relation to prevailing cultural narratives that posit a disenchanted modernity, calling not for the *reenchantment* of modernity but instead proposing an "alter-tale" that accounts for "magical sites already here"—"cultural practices that mark 'the marvelous erupting amid the everyday.'" Bennett makes it clear that this understanding of enchantment is not tied to teleological notions of a divine creator, nor is it solely the province of a "nature" unsullied by human intervention. Rather, hers is an "enchantment without design"—what Timothy Morton (2007) calls "ecology without nature"—in which the artifactual, technological, and electronic are just as capable as the natural at "inspiring wonder, even an energetic love of the world" (Bennett 2001, 9–10). This is not to reduce Bennett's understanding of reenchantment to simply a call to individualize relationships to objects, and in the process reaffirm commodity fetishism. It is to suggest that this "energetic love of the world" may be a more effective grounding for a viable social and environmental justice politics than is the belief in "an alienated existence on a dead planet" (ibid., 4). For Bennett, that it is difficult to feel concern for a disenchanted world means that the affective basis for political action must come from other experiences of and encounters with the world broadly, and industrial materiality specifically.

As I explore here, *Toaster* instantiates this alternate ethics by encouraging us to think of consumer objects as living creatures—a vision it constructs through an emphasis on time and temporality, and specifically, the ways in which processes of obsolescence and disposal within late capitalism work to render objects out of time, out of mind. The film does so by enacting a cinematic version of what cultural anthropologist Igor Kopytoff has termed a "thing biography," in which the practice of recording the shape of a human life over time is applied to the social existence of objects. As Kopytoff (1986, 66-67) explains,

In doing the biography of a thing, one would ask questions similar to those one asks about people: . . . Where does the thing come from and who made it? What has been its career so far, and what do people consider to be an ideal career for such things? What are the recognized "ages" or periods in the thing's "life," and what are the cultural markers for them? How does the thing's use change with its age, and what happens to it when it reaches the end of its usefulness?

Toaster poses and works through many of these same questions in relation to the five appliances in the film, with themes of loss and nostalgia, memory and fantasy, serving as powerful narrative devices (amplified by the kinds of affective attachments and nostalgia that characterizes my relationship to the film as a viewer). The opening scenes of the film, for instance, lead the viewer inside an isolated cottage through a window framed by dilapidated shutters—a subtle detail that sets up the pending story in a temporal context of abandonment or neglect: a time outside of time. The design of the appliances, too, evokes a sense of anachronism; the toaster is the classic, basic model we associate with the streamlined aesthetic of the 1950s, while the melodramatic vocal stylings of the dial-faced radio bring to mind an announcer from radio's golden age. These markers of temporal misplacement, which would have been obvious to audiences even when the film was current, imply that the cottage and its electronic inhabitants are somehow stuck in a time warp, in which nothing has changed for decades. The vacuum cleaner Kirby suggests as much in his response to an inquiry from the electric blanket as to their plans for the day. "What do you mean, what are we going to do today?" he snaps. "The same thing we've done for the past 2,000 days: work!"[2] The implication is that the appliances are stuck in the past because time itself is stuck—like a broken record, it skips—and that time is stuck because the appliances have been abandoned or wasted. The chores they continue to carry out day after day no longer have value because no one is there to need them.

Toaster similarly positions the appliances as outside time in its multiple scenes of fantasy and reminiscence. In one early scene we see Blanky, the electric blanket, fantasize about the day his beloved master will return; although many years have passed, the Master appears within Blanky's fantasy as the child he was when his family last visited the cabin, a living image of the child in the photograph his family left behind. Likewise, Lampy the bedside lamp affectionately recalls a time when the Master replaced his burned out bulb: "I remember the first time my bulb burned out. And I thought, 'That's it! It's over! I'm burned out! 86'ed! To the showers!' But then the master put in a brand new bulb. And I just glowed." Here both retrospection and fantasy—a reflection back and forward in time—act as a preservative, removing their subjects from time or stopping its effects, like the childhood image of the Master frozen in the photograph Blanky clings to.

Yet the film's preoccupation with the passage of time is ultimately less historical than nostalgic. For instance, the design and characterization of the appliances is, on the one hand, historically specific to the postwar consumer boom in United States and yet, on the other hand, ahistorical or anachronistic in that it refers to *no particular* time after World War II, and in many cases jumbles its postwar cultural references with those from the prewar cultural landscape. Early on in the film, the radio belts out Little Richard's "Tutti Frutti" while the appliances go about their daily chores—a scene that suggests the early years of rock and roll in the 1950s. Other scenes, though, include references to Franklin D. Roosevelt and the Depression, the invasion of Normandy, Al Jolson's finale in the 1927 talkie *The Jazz Singer*, Harry Houdini, and Teddy Roosevelt. As such, the appliances, both as temporal markers and in their own memories and fantasies, serve as sites of nostalgia rather than historical referents; the past to which they refer is mythological or ideological as opposed to historically accurate or internally consistent.

In one theorizing of the ideological functions of nostalgia, Arjun Appadurai has argued that in societies where consumption patterns are organized by "fashion"—"a continuous change of small features" that prompts people to replace goods before they're worn out—the cultivation of a longing for the past mystifies and helps sustain such patterns. Nostalgia induces consumption not through "the evocation of a sentiment to which consumers who really have lost something can respond"; as Appadurai (1996, 77–78) observes, it does so by convincing prospective consumers "to miss things they never lost." In the case of *Toaster*, we might say that in asking viewers to identify with obsolete consumer goods—thereby presenting a world in which older is better—nostalgia appeals to an illusory halcyon age of consumption more than it actually critiques the logic of obsolescence that drives modern consumption practices. In short, one could assert that the film *promotes* consumption precisely by seeming to repudiate it, and that this mystification occurs through the cultivation of nostalgia for a world we never actually lost: a world of consumer durables devoted to their masters.

But how can we account for the nostalgia of the appliances themselves, which emerges from an awareness (rather than mystification) of displacement from time? If anything, one could say that consumer goods that feel nostalgic in the face of their own ephemerality bear out the contention undergirding Appadurai's entire exploration of temporality and consumption: that consumption is not "the end of the road for goods and services, a terminus for their social life, a conclusion to some sort of material cycle." According to Appadurai (ibid., 66), such a notion is "an optical illusion," and "in order to get rid of it we need to resituate consumption in time." Likewise, in dramatizing the story of consumer objects postconsumption, *Toaster* denies this same illusion. The appliances' nostalgia is, to that extent, less a desire to displace time than it is for temporal reinsertion—a desire to journey out from the isolated cottage in order to reunite with the Master.

The sort of environmental ethic that emerges from the film's ascription of desire or agency to obsolete commodities becomes clearer in scenes where the film uses reflection in a second, more literal sense. Alongside multiple scenes of looking backward and forward in time are scenes in which various characters apprehend and admire themselves in Toaster's chrome surface. In one scene, a gang of squirrels mesmerized by their reflection in Toaster's casing chase him into the forest, where he escapes by hiding behind a tree-covered hedge. Growing beside the tree is a flower, and as Toaster peers around the hedge to see if the coast is clear, the flower catches a glimpse of itself in Toaster's reflection and embraces him. "It's just a reflection," Toaster tells his would-be suitor gently. "I'm not a flower." Yet the flower persists in its affections, chasing Toaster off once more. A moment later he peers through the hedge; as the leaves part we see the flower dying, its wizened petals dropping to the forest floor. Seemingly a rendering of the myth of Narcissus—in which enchantment with the reflected image results in danger or tragedy—the symbolic use of reflection here could easily be taken as a familiar statement on the clash between nature and technology, where the intrusion of the toaster into "nature" results in the death and destruction of the latter.

Complicating this reading, however, is a second scene that echoes themes of reflection and misrecognition, yet with a crucial difference that suggests the greater proximity of the film's environmental ethic to Bennett's enchantment without design than to Romantic images of a sacrosanct "nature" separate from and disenchanted by industrial technology. In this scene, as the five appliances camp out one night en route to the "city of lights," Toaster has a dream in which he is toasting bread for the Master, who makes faces at himself in Toaster's reflection as he waits. The dream quickly turns into a nightmare when black smoke begins billowing out of Toaster's slots and filling the room, and then forms into a hand that snatches the Master out of Toaster's reach. In his place, a demonic clown garbed in a firefighter's suit appears and instructs Toaster to run; Toaster flees, pursued by a wall of water from the clown's fire hose.

As in the scene with the flower, enchantment with the reflected image results in danger or tragedy, but here the narcissistic mistake has different implications for an environmental ethics of waste and consumption. Unlike in the flower scene, the enchantment to which the Master falls prey hinges not on an overinvestment of reality to the image (taking the reflection of self as a separate other) but instead on an assumption that the other, as reflecting surface or medium, is inanimate—*dis*enchanted, merely a thing and thus incapable of returning the Master's gaze. This is underscored by the familiar trope (seen in *Toy Story* and numerous other animated films) of the machines as apparently inanimate objects hiding their real, living faces from the human characters, around whom they merely pretend to be lifeless and faceless. (*Toy Story* in fact exploits this trope at its climax, with the toys staging their escape from toy torturer Sid by strategically revealing their animate nature in order to terrify him.)

The use of reflection in this latter scene suggests that the dangers of misrecognition are not simply psychodynamic or social but ecological as well. In using the reflective surface of a toaster to stage the narcissistic encounter, the film suggests that obsolescence or fashion, as the cultural logic of consumer capitalism that has been so ecologically destructive, at its core results from a kind of social misrecognition in which we derive social identity from commodities, seeing in them a projected image of our own illusory coherence. Nature here is not a distinct realm opposed, disenchanted, and destroyed by technology but rather *what we fail to see* in the toaster, as an object that has a life in time and geographic space: made somewhere by someone out of materials extracted from the earth that will return to the earth in some form, somewhere. The literal use of reflection in this scene thus indicates that ephemerality and obsolescence arise from a mistaken view of commodities specifically, and materiality more generally, as the self-identical appendage of ego or will, insentient matter that does the bidding of its master. The individualist focus on waste as hated object, or as symbol of nature's disenchantment, is arguably an extension of this misrecognition, diverting attention not just from the systemic nature of environmental problems but also from the more fundamental issue of how we conceptualize and relate to materiality itself within capitalist modernity.

Yet *Toaster*'s environmentalism is never explicit, as its primary ethical concern at all times lies with the effects of overconsumption on inorganic versus organic life. While we could read the dying flower scene as evidence that the film is concerned with a nature "out there" as plants and animals, and whereas later films like *WALL-E* are primarily concerned with the fate of a nature reified as plant life, *Toaster*'s environmental ethic is arguably closer to Bennett's enchantment without design or Morton's ecology without nature, challenging our expectations about what the environment is (and hence the proper object of our ethical attention as environmentalists).

The implicitly ecological dimensions of the film's thing biography become clearer toward the end of the movie, when after arriving in the city of lights, the appliances locate the apartment where the teenage Master now lives with his mother. As they discover, it is an urbane, tasteful space furnished with the latest models of all the technological objects that signal affluence—big-ticket items like computers, stereos, and home entertainment centers, but also the microwaves, egg beaters, and floor lamps that represent the "inconspicuous consumption" (Shove 2003) that constitutes the baseline of expectations for a modern standard of living within the Global North. The film's emphasis on the new appliances' contemporaneity (which by now appears laughably anachronistic) suggests that the older appliances' arrival is in many ways a confrontation between the postwar era and Reagan years—two historic periods of consumer spending in the United States. Technological obsolescence here is therefore literalized as a battle between new and old machines; greeted by cutting-edge doppelgangers who are none too delighted by their arrival, the five appliances quickly realize that they are out of time, not just because they have been abandoned and replaced, but because they are at the end of their useful social lives. "More, more, more! Everything you wanted and more!" the new appliances boast in song, before booting their visitors from the Master's apartment window into the dumpster below, where the older appliances are carted off to the city dump on the city's outskirts.

This scene is doubly interesting in that it juxtaposes shots of each appliance's disposal with shots of the Master searching for that appliance at the cottage, having ironically driven there to retrieve his old stuff to take with him to college just as the appliances arrive at his apartment to find *him*. "Where's the toaster?" he exclaims in frustration—as back in the city, Toaster falls out of the Master's bedroom window and lands in the dumpster. Time has become disjointed; they are all too late. The very moment the Master attempts to apprehend Toaster is the moment Toaster disappears altogether from the category of the socially perceptible. This is to say, the very moment the appliances reinsert themselves into time is the moment they realize their time is up.

The climax of the movie thus revolves around the appliances' attempt to solicit the Master's apprehension before their discovery at the dump by an angry-faced magnet whose job it is to transport them, as waste, to the trash compactor.[3] After skillful and subliminal maneuvering on the part of a black-and-white television in the Master's bedroom—an acquaintance of the appliances from the cottage days—the Master drives out to the dump to search for secondhand items to replace the ones he was expecting to find at the cabin. Yet even after arriving he cannot see what he has unintentionally come for, and continually misses the appliances among the piles of junk and refuse.[4] Although they attempt to intercept his line of vision by laying themselves out conspicuously, the Master always manages to look elsewhere or look away at the last second. It is not until the Master by chance finds the childhood picture of himself from the cottage, dropped by Blanky en route to the trash compactor—that the appliances become visible to him, for in that moment he sees himself as the appliances see him. Rather than misrecognizing himself in the reflection of commodities, he recognizes himself in their desire, and it is only then that he finds the appliances where in fact they have always been.

How to See Stuff: Two Perspectives

If biography can be used to tell the story of a thing's movement through time, then the biography that *Toaster* records charts the appliances' temporal journey from use to obsolescence to disposal to reclamation and reuse. Movement in time is thus a movement in and out of the category of waste, as a cycle of valuation that also subjects the appliances to different conditions and thresholds of social visibility. Central to the film's emphasis on temporality, then, are the politics of perception—processes of reflection, misrecognition, recognition, and visibility. But why would a politics of perception be key to the process of resituating consumption in time? Why do the appliances apprehend the Master if he cannot readily apprehend them, and what exactly does he fail to apprehend? Why are the lives and deaths of objects secret—not just as represented in film, but also actually, materially, within conditions of commodity fetishism—and what are the ecological implications of making perceptible this secret?

Both environmental and cultural studies have something to say about these questions, given that both fields include critical genealogies in which the revisioning of commodities becomes a central means of critiquing the system that has produced and discarded them. Yet to understand the connections between these seemingly disparate fields, it is first necessary to explore why commodities are out of time in an economic sense. Why is consumption the condition of temporal and hence cognitive displacement? That is, why are we generally not aware of the story of the things we consume, such as where they come from before they get to us and where they go after we finish with them?

One possibility, as Kopytoff (1986, 69) has pointed out, is that commodities present the appearance of having a uniform history: "To use an appropriately loaded even if archaic term, to be saleable or widely exchangeable is to be 'common'—the opposite of being uncommon, uncomparable, unique, singular, and therefore not exchangeable for anything else." This distinction between singularities and commodities is an extension of a more fundamental one between persons and things, which Kopytoff states has been conceptually central to Western modernity. Like persons, singular objects have personal or individual values, and thus a particular tale to tell, whereas the common value of commodities gives them the appearance of having no story (or no particular story worth knowing).[5]

Still another reason behind the temporal displacement of commodities is what ecological economist Thomas Princen (2002) has called distancing. As I have previously explored in relation to urban industrial methods of wastewater treatment (Cortez 2005, 104), distancing refers to the ways in which production and consumption decisions become separated within global markets, leading to the material and cognitive externalization of economic costs and consequences. According to Princen, we become incapable of assessing or even perceiving these costs because the very scale of market relations breaks down the feedback loops that signal the embeddedness of economic agents within ecological systems. From the extraction of raw materials to their processing, purchase, consumption, and disposal, distancing means that we make economic choices without knowing what their impacts are on land or other people. Distancing, then, precludes any easy perception of an object's historicity or futurity; it is an epistemological isolation from materiality in which all that is "visible" is the moment of exchange and consumption.

To illustrate this idea, Princen compares the international food trade to a local farmers' market. In a chain supermarket, shoppers make decisions with a near-complete lack of knowledge of the social or ecological conditions of production. Princen (2002, 119, 118) explains that

> when I buy grapes from Chile in my North American supermarket, I know nothing about the grower and I am never likely to. . . . As a result, I have no way of knowing if my consumption is supporting or undermining that farmer, economically or ecologically. With no feedback or with uncertain feedback I can only assume and, as with most of us, prefer to assume, that my purchases are supportive. . . . In the local market . . . I know something about local farmers. I have a pretty good idea of who they are, what they value, and where they fit in my society. I can certainly see them face-to-face. . . . When I ask about the produce, how it was grown, how to cook it, and so forth, I am likely to get a straight answer."

While this is perhaps a romantic view of farmers' markets, Princen's example also provides a useful heuristic for thinking about social forces of distancing and how they have impacted human relations to the material world, whether animate or inanimate. In the first instance, produce is a commodity in Kopytoff's sense of the term: as an exchangeable item, only price distinguishes it from Cheerios, pencils, or computers; as an object it is effectively faceless. And as Princen points out, price is the story of no story—a poor indicator of whether the produce it describes was produced sustainably, or whether pesticides used in its production polluted nearby waterways or sickened fieldworkers. While produce in a farmers' market is still a commodity, it is arguably more of a singularity in that one can more easily imagine its life by attaining information as to its *particular* history of production.

Environmentalists have long critiqued the forces of distancing responsible for environmental degradation and social injustice by resituating consumption in time, revisioning commodities as singularities with particular histories to be traced. Since the 1960s, such an approach has been integral to environmental assessments of industrial production in the form of life cycle assessment. Often referred to as cradle-to-grave analysis, life cycle assessment is a method for measuring the total environmental impact of a product over the entire course of its production, consumption, and disposal (United Nations Environmental Programme 2001, para. 1). While environmental analysts have generally applied life cycle analysis to the production and distribution of objects, it has been used to understand the impact of consumption as well.

An excellent example of this is John C. Ryan and Alan Thein Durning's *Stuff: The Secret Lives of Everyday Things*, which presents life cycle analysis as a critical exercise for making visible one's individual implication as a consumer within global economic systems—an exercise that effectively undoes forces of distancing. In *Stuff*, Ryan and Durning (1997, 4) "trace the wake" of everyday objects encountered in an average day by "a fictional, typical North American—a middle class resident of Seattle," revisioning these objects in terms of the places, people, materials, labor, and socioenvironmental impact they contain yet obscure. In their prologue to the book, the authors introduce the reader to this fictional-but-typical consumer, whose chance insight one day as she is cleaning out her basement provides the impetus for the life cycle analysis to come:

There I was, piling old paint cans into a cardboard box when something caught my eye. It was a sticker that had fallen off the back of who-knows-what stowed in the basement. It said, "Made in Taiwan." I'd seen thousands of such stickers in my life without ever giving them a second thought. Taiwan. Taiwan. Not just a word on a sticker. It's an island. A country. A real place with real people across an ocean from me. Suddenly, the overloaded shelves around me looked differently. I was stripped of the illusion that stuff comes from stores and is carted "away" by garbage trucks: everything on those shelves came from a real place on the Earth and will go to some other place when I'm done with it. Everything had a history—a trail of causes and effects—and a future. Everything had a life, of sorts. (ibid., 5)

In this way, Ryan and Durning resituate consumption in time by revisioning commodities as singular and animate—where did *this particular* hamburger, shirt, or bicycle come from, and who did it touch along the way?

The political possibilities that lie in revisioning commodities, rather than simply despising and displacing waste, is equally the province of a certain strand of cultural studies, especially in the work of scholars interested in applying Walter Benjamin's philosophy of history to the cultural analysis of consumption. A literary critic, translator, and art historian, Benjamin wrote and published during the 1920s and 1930s as part of a network of German Jewish intellectuals known as the Frankfurt school. Writing during the buildup and aftermath of the Third Reich, the key theoretical contribution of the Frankfurt school was an analysis of modernity and modern culture as beset by a fundamental contradiction between progress and catastrophe. This contradiction is well expressed in a famous quote from one of Benjamin's last pieces, written not long before his suicide while trying to escape the Nazi occupation of France. In this quote, Benjamin references an image from the expressionist painter Paul Klee's *Angelus Novus*, using this image to articulate his critique of a progressive history, in which "the past exists [only] to confirm the superiority of the present" (quoted in Neville and Villenueve 2002, 18):

A Klee painting named "Angelus Novus" shows an angel looking as though he is about to move away from something he is fixedly contemplating. His eyes are staring, his mouth is open, his wings are spread. This is how one pictures the angel of history. His face is turned toward the past. Where we perceive a chain of events, he sees one single catastrophe which keeps piling wreckage upon wreckage and hurls it in front of his feet. The angel would like to stay, awaken the dead, and make whole what has been smashed. But a storm is blowing in from Paradise; it has got caught in his wings with such violence that the angel can no longer close them. This storm irresistibly propels him into the future to which his back is turned, while the pile of debris before him grows skyward. This storm is what we call progress. (Benjamin 1968, 257)

For Benjamin along with his Frankfurt school colleagues, modernity is characterized by this contradiction between progress and catastrophe, represented metaphorically as an accumulation of waste—for every claim to scientific advancement, one had the historical realities of the death camps; for every modern skyscraper, one had the modern drive to destroy nature. Benjamin's colleagues extended this critique to the post–World War II rise of mass culture as well. Radio, television, and film purported to entertain and comfort, even as they disguised their real function as tools of conformism and fascism—what the Frankfurt school referred to as the rise of the "culture industry." In this analysis, mass culture itself was waste, "expendable to the highest degree, superfluous, trash" (quoted in Neville and Villenueve 2002, 19).

As is evident in the quote above, Benjamin shared the critique of history as progress, yet he maintained different ideas about how to engage with waste as a symbol of the contradictions of modernity evident in its cultural objects. In recent decades, this had led many cultural studies scholars to turn to Benjamin in their efforts to write about the politics of consumption in late capitalist societies, particularly in their attempt to complicate straightforward equations of commodity culture with domination, and its enjoyment with deception or false consciousness. As Susan Buck-Morss (1989, ix) states, Benjamin's critique of culture within capitalist modernity centers not on a rejection of mass culture as rubbish or waste but instead on a "dialectics of seeing" that "[takes] seriously the debris of mass culture as the source of philosophical truth." In this reading, the Benjaminian take on waste focuses not on its absolute negativity but rather on the utopian prospects that lie hidden within the discarded consumer items of a progressive history, and the possibility that a mode of perception attuned to these prospects can create an enchanted present.

Thus, for instance, Angela McRobbie in "The Place of Walter Benjamin in Cultural Studies" examines the implications of waste in Benjamin's *Passagen-Werk* (Arcades Project), the unfinished assemblage of notes on the Paris arcades—shop-lined boulevards canopied by glass structures that in many ways anticipated today's indoor shopping malls. As McRobbie writes, the arcades were outdated even in Benjamin's time, and for him they became a symbol of modernity itself, the symbol of a history best legible in its ruins: once proof of the present's superiority, now discarded and worthless. In McRobbie's reading (1993, 93), Benjamin's fascinated encounter with the Paris arcades opened him to a central insight: if the objects of industrial culture had the ability to "enchant and be forgotten" in the act of consumption, they might also contain "another mode of enchantment which occurred in and through these same objects and which in the process of being revealed and remembered could play an active role in the re-enchantment of society." Where the more traditional materialism of his Frankfurt school colleagues saw consumer culture as proof of the irrationality of rational systems—the contradictions of a history that at every moment claimed progress even as it left catastrophe in its wake— Benjamin instead revisioned these objects as a site of revitalization, the point at which possibility emerged precisely from these contradictions.

It is easy to see how *Toaster* brings together these two genealogies of revisioning, and in the process suggests an environmental ethic of enchantment grounded in a perceptual mode attuned to the historical lives of objects. This way of seeing does not fetishize commodities so much as, like Benjamin's *Passagen-Werk*, "take materialism so seriously that the historical phenomena [embedded in them are] themselves . . . brought to speech" (Buck-Morss 1989, 4). In its vision of objects that speak their own particular histories (and futures), *Toaster* quite clearly blurs Kopytoff's distinction between persons and things, depicting objects that are also persons, hybrid forms whose singularity commands ethical regard even as their commonness renders them always potentially commoditizable (such as in the scene in which the appliances end up in a parts resale shop, depicted as a gothic chamber of horrors). The film thus presents what according to Kopytoff would be a Western conceptual paradox: the idea that mere things are deserving of ethical regard, and for that reason should not be thrown away before the end of their useful life in order to acquire more. In this context, the figure of the master does not suggest a paternalistic relationship to nature or materiality so much as it does the nonrelationship of consumption; mastery is the inability to apprehend the imminent singularity of objects designed for human exchange, and hence their secret lives: the hidden history of where they come from, where they end up, and who they affect in the process of this journey. The film implies that if human masters *could* apprehend this life, they would in effect no longer be masters. Relations to materiality would be grounded instead in mutual recognition and desire: the "joyful attachment" highlighted by Bennett (2001, 12), or what one might call *object love*.

Object Love and Its Limits

What would it actually look like to reciprocate the desire of a machine—to be joyfully attached to it, to love it? Taken literally, this would seem to risk advocating or even celebrating the same kind of individualist waste ethos I have been arguing against, and to which I have claimed *Toaster* suggests an alternative.[6]

The question looms large for me, as someone who has been as deeply involved in environmental justice organizing as I have been in the critical analysis of popular culture. What is the utility of joyful attachment in the case of a military base whose sloppy disposal of degreasing chemicals has contaminated the groundwater of neighboring communities, causing high rates of amyotrophic lateral sclerosis? Or in the case of a city government whose progrowth policies spell a commitment to continued investment in nuclear energy, despite the lack of a solution to the nuclear waste problem, even as accumulating news reports from postearthquake Japan foretell the worst nuclear disaster since Chernobyl? What is the relevance of object love this very morning, as I help clean up the trash that has collected at the edge of the wetlands on the campus of a tribal university—an act of pride as well as defiance against a proposal by the state transportation department to route a highway expansion through the wetlands? We take photos when we are done, smiling in orange safety vests behind a mountain of black plastic bags filled with all manner of garbage and debris: plastic water bottles, beer cans, cigarette butts, hubcaps, deflated basketballs, endless Styrofoam. The disintegrating Doritos bag I pick up makes me sad and angry, not sympathetic, when its metallic sheen flakes into too many slivers to collect and throw away.

Similarly, the ecological ethic of *Toaster* looks quite different when we consider the film as material object rather than narrative. I can no longer watch the copy of *Toaster* I used to write this analysis, because I no longer have a VCR, having finally abandoned cassettes for DVDs or streaming video on my laptop. What happened to that VCR—functional but no longer valuable—after it left my hands, and what is the environmental fate of the cassettes that remain, equally functional yet similarly useless? What happens when we think of cherished narratives as disposable plastic objects? What does the obsolescence of media formats tell us about the limits of object love? Can this really be a tenable basis for an alternative environmental ethic?

Hawkins herself acknowledges the difficulties of sustaining a call for revisioning rather than despising the nonorganic, frequently toxic kinds of by-products produced by commodity cultures, and is much more adept at suggesting what an ecological ethic of pleasure or desire might look like when discussing people's careful relations to recyclables and worm farms, or a Brisbane composting education campaign called Enjoy Your Garbage. Waste objects are easier to love when their excess can be imagined as generous and abundant, and their care envisioned as a pleasurable form of husbandry that also cultivates an ethical or aware self.

It would be a mistake, however, to read a call to revision waste as a call for its transvaluation—a simple movement in all cases from condemnation to celebration—given that the context for this call is the very destructive impacts commodity cultures have on people and the planet. Object love, then, is ultimately less an ascription of positivity or an attachment to particular objects than it is a sensuous recognition of temporal and geographic embeddedness, an apprehension of the immanent singularity of objects that prompts a revisioning of their lives in time and space, and hence the systems producing and disposing of them. Here, love is a quality of attention, a willingness to consider and be moved, affectively and thus politically. As per Bennett, this willingness challenges narratives of modernity as disenchanted by positing an enchantment already here, accessible via an affective openness or sense of wonder. How *is* it that individual habits and practices of energy consumption are linked to a network of inconspicuous consumption that begins with the extraction of uranium from the earth, and ends with Yucca Mountain or Chernobyl? How *is* it that the production and consumption of petroleum-based materials, which makes possible the circulation of media narratives as much as the Styrofoam cup crumbling in the wetlands, has come to organize modern economies?

It is in this sense that I would argue there is something ethically defensible, even ethically compelling, about the regard for or attention to materiality as *things* imagined in *The Brave Little Toaster*. For if, as historian Susan Strasser (1999, 11) has argued in *Waste and Want*, mass production effected the loss of a "kinesthetic knowledge of materials"—a knowledge of how things are put together, and therefore how to repair or reuse them—as well as the sort of attachment to objects that once discouraged their disposal, an object love predicated on seeing things differently might provide a much better corrective to epistemological distance than the standard ecoinjunctions to consume less or forgo desire.

In fact, it is because normative encounters with waste objects often reinscribe the individualizing focus they aim to challenge—Where did that mountain of trash go after we dredged it from the wetlands? Away: that was all that mattered—that Hawkins calls for an environmental ethic of affect and sensuous attachment. For as Hawkins (2006, (31–33) points out, environmental ethics generally are too often grounded in guilt, duty, and obligation, and our subjective engagements with waste in particular are structured by "legislated and normative moralities" operationalized through "a monitoring and disciplining relation to the self." The problem with this, Hawkins (ibid., 34–39) contends, is that such an ethic "denies other ways of being with waste," and with objects of consumption more generally:

> There is no room for disgust or horror or pleasure or resentment. No room for movement between different registers of subjectivity or for any recognition of how changes in practices of the self may affect other ethical sensibilities. . . . So much waste education secures an obligation to new [less environmentally destructive] techniques and habits by insisting that the threat of waste be mastered: reduced, reused, recycled. This fantasy of control establishes the responsible self as separate from the world. It maintains an absolute alterity of waste and blinds us to an awareness of how our relations with it are fundamental to the very possibility of life.

Again, this is not to deny the reality that there are contexts in which it makes political sense to think of waste as other and separate from the self—as in many community struggles around the politics of siting and contamination. But it is to suggest that what theorists like Hawkins and films like *Toaster* do is ask us to attend to the *affective* basis for environmental action. Revisioning is less about what we do with waste, as the object of policy and activism, than with what happens when we allow ourselves to encounter it, and with how these moments of attention—the shock, wonder, delight, or disgust waste provokes—might serve as subjective and micropolitical ripples that fan outward to create more environmentally responsible structures, practices, and technologies. As to whether this encounter is enough, there is no guarantee. According to Hawkins (ibid., 131), "All that can be argued is that these moments engage us in different relations with commodities that could be a source of more self-conscious material practices."

Undoubtedly, reading the ethics of *Toaster* as narrative doesn't change the way the film is produced or disposed of as a material object, nor the social and environmental consequences of these practices. And yet, might not such a reading surprise us into an awareness of waste along with the materiality of its temporality and sociality in ways that previously remained beyond perception? More broadly, where *would* more systemic, pragmatic changes in what we do with waste come from, if not from these affective encounters with materiality as animate, as worthy of ethical consideration?

It makes sense, then, to ask what kinds of affects are produced by images of obsolescence in popular media, and what kinds of subjects are made possible—or inhibited—by these affects. In particular, what should we make of the recent subset of animated narratives within children's culture, which in lamenting obsolescence, seem to run at cross-purposes with what Shirley R. Steinberg and Joe L. Kincheloe (2004) have dubbed "kinderculture," or the increasingly corporatized construction of childhood around consumption imperatives? Viewed cynically, one could easily see these films' waste sympathies as an obfuscation of the ways in which popular film as media form or material object participates in the ruthless logic of commodification and obsolescence it seems to decry on a narrative level. In this reading, an enchantment with obsolete toys, robots, or toasters promotes attachment to individual objects, but without a parallel attention to the network of temporal and geographic relations in which these objects are embedded.

The question of whether and how recent ecoconcerns with obsolescence in animated film celebrates and/or condemns consumption is a larger project that I can only begin here. Undoubtedly, a film like *Toaster* must be examined in relation to its thematic descendants in the *Toy Story* series and *WALL-E*, and the broader historical concern with human-thing relations in animated film. These must be in turn situated in relation to questions of the environmentality of film production and distribution, subjecting technologies of popular media circulation to the same life cycle analysis that the obsolescence of films as narratives seem to call for. For the purposes of this chapter, though—more narrowly focused on the kinds of ethical or political subject constructed in *Toaster*'s affect of attachment—it perhaps suffices to ask, Does sympathy for waste in this film construct a subject who "forgets" the materiality beyond narrative, or who is driven to uncover, recover, or revision it?

The conclusion of *Toaster* does seem to affirm the necessary connections between modes of perception and practices, or an ethical encounter with objects as singular and a broader pragmatic resistance to the cultural logic of disposability. "Really now," the Master's girlfriend Chris remarks as the Master repairs Toaster after the latter's plunge into the jaws of obsolescence. "Why don't you just go out and buy a new one?" "Are you kidding?" Rob, the Master, responds. "Where could I find another toaster like *this*?" He's joking; the camera closes in on a shot of a badly mangled Toaster. "Like *that*?" replies a skeptical Chris. "Probably nowhere."

There is, arguably, a kind of attachment, a nostalgia, a celebration of the singularity of objects that closes off knowledge of structures, systems, and histories, that is politically naive. But there is also an attention to particularity that might open up epistemological space for this knowledge; there is an attachment to and enchantment with a story that raises questions about the narrativity of all materiality along with the temporal and spatial life journeys traced by its particular objects. There is the danger, too, that in the end, one form of attachment cannot be so neatly bracketed off from another, object love from its limits, the individualizing vision of enchanted things as reflective of commodity fetishism and the temporalizing attention to particularity that disrupts it. I recognize that I run the risk of reinscribing what I want to undermine. But it is a risk I want to take nonetheless, because of what is at stake for environmental justice struggles in our ability to see and think about the impacts and effects that global economic forces externalize from sight and mind. Where they reappear to us, cultural images of waste and the affects they produce provide just such an object lesson.

Notes

1. All dialogue in the text is cited from the VHS release of *The Brave Little Toaster* (Buena Vista Home Entertainment, n.d.).

2. In this scene, the final death of objects as waste is imagined as the reduction of individual machines with particular life histories into abstract and identical cubits of metal. *WALL-E* conceptualizes object death in similar terms, in the scene where WALL-E and EVE get thrown down the trash chute of the ship where an evacuated humanity has lived for seven hundred years. Landing in the equivalent of the ship's dump, the two robots narrowly escape death at the hands of a

compactor aimed at reducing them, along with all kinds of other refuse, to cubes of compressed material—a final heterogeneity that nonetheless embodies a high entropy state of uselessness.

3. Added to the poignancy and intrigue of this scene for me is my awareness, in watching the film as an adult, of my own location in and movement through time as history: the sort of publicly accessible city dump featured at the film's climax, and to which I remember going as a child with my grandparents, is no more, for good reason. As environmental historian Martin V. Melosi (2005, 213–214) documents in *Garbage in the Cities: Refuse, Reform, and the Environment*, the passage in 1976 of the Resource Conservation and Recovery Act mandated "the development of environmentally sound disposal methods and the protection of groundwater with new landfill standards," which effectively required the closing of open dumps throughout the 1980s.

4. The person/thing distinction has been used as justification for oppressive practices historically central to modernity (for example, slavery as an extension of colonialism broadly), but is also central to the moral basis for materialist political projects that resist or critique domination as "thingification." For instance, Jennifer Daryl Slack and Laurie Anne Whitt (1992, 573) see cultural studies as resting on a commitment to the idea that "human beings—whether as a species or as individuals—are intrinsically valuable and enjoy moral standing, which must be respected and reflected in how they are treated; and . . . that oppressive social and political formations are objectionable and to be resisted to the extent that they are indifferent to this." This is not to imply that because the person/thing distinction is ontologically essential to modernity, we should dispense with modern critiques of domination as instrumentalization. Rather, it is to argue (following from Hawkins) that questioning the givenness of subordinating things to persons might in fact accomplish this more effectively.

5. For instance, as a graduate student teaching assistant several years ago, I once led discussion sections for an undergraduate course on technology and the environment. The students in this class were asked to read Ryan and Durning's *Stuff*, and write their own object biography, including at the end of their essays a discussion of possible solutions to the social and environmental costs their biography uncovered. After tracing the wake of his desktop computer, and in the process uncovering the effects of e-waste on workers and the environment, one student memorably suggested that one possible solution lay in loving one's computer enough that one never wanted to throw it away. As he explained, each computer system that entered into his possession over the years had acquired a special significance for him, serving as the repository for particular memories and experiences. He thus never discarded his old systems when he acquired new ones, and in this way, in his attachment to specific machines, had avoided contributing to the e-waste problem.

References

Appadurai, Arjun. 1996. *Modernity at Large: Cultural Dimensions and Globalization*. Minneapolis: University of Minnesota Press.

Bennett, Jane. 2001. *The Enchantment of Modern Life: Attachments, Crossings, and Ethics*. Princeton, NJ: Princeton University Press.

Benjamin, Walter. 1968. Theses on the Philosophy of History. In *Illuminations*, ed. Hannah Arendt, 253–264. New York: Harcourt Brace Jovanovich.

The Brave Little Toaster. 1987. Dir. Jerry Rees. Videocassette. Burbank, CA: Buena Vista Home Video.

Buck-Morss, Susan. 1989. *The Dialectics of Seeing: Walter Benjamin and the Arcades Project*. Cambridge, MA: MIT Press.

Cortez, Marisol. 2005. Brown Meets Green: The Political Fecology of PoopReport.com. *Reconstruction* 5 (2). http://reconstruction.eserver.org/052/cortez.shtml.

Hawkins, Gay. 2006. *The Ethics of Waste: How We Relate to Rubbish*. Lanham, MD: Rowman and Littlefield.

Kopytoff, Igor. 1986. The Cultural Biography of Things: Commoditization as Process. In *The Social Life of Things: Commodities in Cultural Perspective*, ed. Arjun Appadurai, 64–94. Cambridge: Cambridge University Press.

Marx, Karl. 1978. Capital, Volume One. In *The Marx-Engels Reader*, ed. Robert C. Tucker, 294–438. 2nd ed. New York: W. W. Norton and Company.

Melosi, Martin V. 2005. *Garbage in the Cities: Refuse, Reform, and the Environment*. Pittsburgh: University of Pittsburgh Press.

McRobbie, Angela. 1993. The Place of Walter Benjamin in Cultural Studies. In *The Cultural Studies Reader*, ed. Simon Durning, 77–96. New York: Routledge.

Morton, Timothy. 2007. *Ecology without Nature: Rethinking Environmental Aesthetics*. Cambridge, MA: Harvard University Press.

Neville, Brian, and Johanne Villeneuve. 2002. Introduction:In Lieu of Waste. In *Waste-Site Stories: The Recycling of Memory*, ed. Brian Neville and Johanne Villeneuve, 1–25. Albany: State University of New York Press.

Paik, Karen, and Leslie Iwerks. 2007. *To Infinity and Beyond: The Story of Pixar Animation Studios*. San Francisco: Chronicle Books.

Princen, Thomas. 2002. Distancing: Consumption and the Severance of Feedback. In *Confronting Consumption*, ed. Thomas Princen, Michael Maniates, and Ken Conca. Cambridge, MA: MIT Press.

Ryan, John C., and Alan Thein Durning. 1997. *Stuff: The Secret Lives of Everyday Things*. Seattle: Northwest Environmental Watch.

Schweiger, Daniel. 2004. *Liner Notes to* The Brave Little Toaster: *Original Motion Picture Soundtrack*. Percepto Records.

Shove, Elizabeth. 2003. *Comfort, Cleanliness, and Convenience: The Social Organization of Normality*. New York: Berg.

Slack, Jennifer Daryl, and Laurie Anne Whitt. 1992. Ethics and Cultural Studies. In *Cultural Studies*, ed. Lawrence Grossberg, Cary Nelson, and Paula A. Treichler, 571–592. New York: Routledge.

Steinberg, Shirley R., and Joe L. Kincheloe. 2004. *Kinderculture: The Corporate Construction of Childhood*. Boulder, CO: Westview Press.

Strand, Ginger. 2008. The Crying Indian: How an Environmental Icon Helped Sell Cans—and Sell Out Environmentalism. *Orion*, November–December. http://www.orionmagazine.org/index.php/articles/article/3642/.

Strasser, Susan. 1999. *Waste and Want: A Social History of Trash*. New York: Henry Holt and Company.

United Nations Environmental Programme. 2001. Life Cycle Analysis. *Sustainable Agri-Food Production and Consumption Forum*. http://www.agrifood-forum.net/practices/lca.asp. Accessed March 14, 2011.

Conclusion: Object Lessons

Stephanie Foote and Elizabeth Mazzolini

Each of the works collected in this volume confirms the insight articulated by John Scanlan (2005, 14–15), who writes in *On Garbage* that "the act of conceptualizing garbage actually transforms it into something else." By meticulously sifting through the various material and symbolic work done by garbage, the chapters reimagine what culture looks like when its most despised category becomes central and foundational. Each of the contributors has, in the process of historicizing and examining the categories of garbage and waste, transformed them into both objects of inquiry and the grounds from which to illuminate how other key historical and social categories—place, personhood, class, agency, and gender, to take only a few examples—are shaped by them.

Throughout this volume, chapter by chapter and as a whole, we have thus tried to show that the act of conceptualizing garbage is also always an act of defining the cultures that have generated it as well as revealing the fault lines that run through any unified notion of what counts for them as important, productive, valuable, and desirable. The chapters tell stories that have too often gone unheard about who has had the power to define what garbage is, how it has been used to classify people and communities, how it has been critical in mapping real and imaginative spaces, and perhaps most important, how it has become the sign of the uneasy coexistence of economic abundance, on the one hand, and environmental degradation, on the other hand, precisely because waste is as much a problem of production as it is one of consumption.

And yet all the contributors have also paid close attention to the shifting definitions of garbage and waste in different moments, and the multiplicity of narratives generated by any single definition of garbage. Municipalities, for example, can see sewage as profitable fertilizer rather than just sludge, and citizens can recast toxic waste as a powerful political platform to describe the invisibility of ordinary life in the face of corporate greed. All the contributors have seen not just how conceptualizing garbage transforms it into something else. They also have argued that paying attention to garbage can itself transform an individual subject as well as how we conceptualize culture in relationship to the circulation of different kinds of real and metaphoric garbage. As a whole, the chapters have shown how many narratives about garbage and waste our culture asks us to hold in suspension or repress at any given moment. The chapters make it clear that as citizens, consumers, activists, and producers, we are asked to generate and yet separate ourselves from garbage; we recognize it as perfectly ordinary when we see litter on the street and ignore it when it is the invisible by-product of a steady job at a physical plant; and we disbelieve as well as guard against its ability to cause human illness and misery.

Indeed, as we compiled this volume, narratives about garbage and waste seemed to collect nearly as speedily and voluminously as trash itself. The work of conceptualizing and narrating garbage is not merely academic, and not merely the province of academics or activists. It is central to daily life along with the habits of existence, and shapes the way we see the world around us. Garbage has been rendered highly visible (as well as compelling and strange) in the recent spate of television programs that conceptualize it as a material and social category. For every news story about landfills or toxic waste dumps, for every activist who works in the medium of garbage, or who wants to transform garbage into fuel at a landfill, there are other more audience-pleasing stories about accumulations of stuff, about treasures mistaken for junk, about ordinary suburban houses that on the inside look like landfills. Such images are at the heart of popular forms of entertainment on television—programs that focus almost entirely on people's relationship to garbage and waste. Even as we engaged the academic archive of scholarly work on garbage, we came to realize how many popular stories about individual relationships with or responses to garbage now exist. The emergence of garbage as the engine of popular entertainment reveals a fascination with it that seems to include revulsion as much as it does awareness.

Take one case study. In August 2009, the US cable network A&E (the Arts and Entertainment Network) began airing a program called *Hoarders*, which has since become one of its most popular features. Each hour-long episode concentrates on one or two households dominated by a hoarder, and spotlights extensive interviews with the designated hoarder, their friends and family members, and mental health professionals. As with any genre, the story told by any episode of *Hoarders* is standard—somehow, the hoarder confesses, objects have just overtaken the household, as if they have a kind of agency or will that cannot be stopped. Certainly every person who is identified by the episode's producers (frequently at the instigation of family members who are crowded out of the house) as a hoarder has a narrative about the world of stuff in which they live. These confessions are indeed what give the therapists provided by the show a handle on how to engage each particular "case." Generally, though, the most loving, detailed, and crucial narratives are not about the mass of stuff, nor are they about the hoarders and their lives. The most detailed and personal narratives are about individual things.

When presented with a seemingly random object, plucked by a professional organizer out of a towering pile of junk, the hoarder in question can describe where it came from, why it was important, what it once meant, and how it must stay in the house after all. Often, the hoarder can portray individual objects in practical as well as abstract terms, sometimes referencing conventional forms of economic value, and sometimes indicating a value that cannot quite be captured by consumer culture, but only by some interior economy that the therapist (or professional organizer) must try to translate into more mundane terms. The program's drama is not just about the titillation it gives to viewers as they look inside someone else's apparent garbage dump but rather about the different kinds of value that are in tension at all points in the show. What is this object worth, and who gets to decide what counts as worth or value? Whose narrative about what is and is not junk will count?

That narrative is critical for the show's success, for *Hoarders* is invested in a metanarrative about how garbage both contaminates and defines people. This is a narrative about personal failing, filth, loss of control, and shame. How, the show seems to demand, can people live in houses that are so dirty? How can they restrict their lives to just a single room and pile the others from floor to ceiling with piles of junk? Didn't they wonder where their cats went? Don't they love their children? How can they have lost control? Why don't they see their lives as we would see them? What does it mean that they have so much, even in the land of plenty, that they are for all practical purposes immobilized? The narrative is completely focused on the pathology of the individual hoarder as opposed to the cultures of accumulation, pathology, and voyeurism that underwrite the show.

If *Hoarders* seems as though it is a singularity, it is not. It has close cousins, like *Hoarding: Buried Alive* and the even more sensational *Animal Hoarders*. Certainly, we can understand these shows as part of a wider reality television programming, with its narrative arc in which a subject is exposed, confesses, and finally gives himself over to a regime of constant vigilance for his own sake. But there is, we believe, a more richly textured political reading of these visual and cultural narratives of such shows. They belong to a rather specific niche of "reality TV" that is dedicated to exploring individual relationships to things and objects, how we classify and categorize them, how we route our identities through them, how we fantasize our political commitments through them, and how we sacralize or condemn them. They are, in short, programs about piles of stuff and the stories we tell about them.

If we focus on *Hoarders* again for a moment, we see that one of the cathartic experiences of every episode—the moment when all the stuff gets cleared out of the house—is actually rather strange. As the army of professional organizers pick up each object and ask the hoarders to categorize it into piles—one pile for things a person really needs, one for things they need to think more about, one pile for things to throw away, and one for things that can be donated or recycled—the audience sees something clearly: it's all junk. What gets thrown away, kept, recycled, or donated to charities—none of those new resting places for the stuff really matters at all. At the moment of viewing, at the moment of decision, it's all garbage; none of it is better or worse, and none of it is "really" valuable at all. The episodes are fables of uncontrolled abundance as well as a lesson in how capitalism pathologically converts desire into necessity, consumers into hoarders, and homes into trash heaps. On the one hand, we think, it seems useful to bring garbage front and center to a more public awareness; these programs, though, seem to be about exceptional relationships to garbage, and therefore may work to keep garbage and waste relegated to the realm of the exotic.

And yet other less pathologizing narratives validate the transformation of both the subject and object under consideration. Some stuff can redeem us, if we only collect and value it correctly. *American Pickers* and *Pawn Stars*, which airs on the History Channel, or PBS's *Antiques Roadshow* present a world in which one person's trash, to recycle the old chestnut, will be another's treasure. The hoarders' task is to recognize that they have become a subject of their possessions and can only attain freedom once they have gotten rid of them. The task of the people who appear on *American Pickers* or *Antiques Roadshow* might appear quite different, but it is part of the same continuum of learning to revalue and renarrate the objects that surround us. Here, the pure and absolute value of objects must be assessed in some fantastical world in which humans do not obscure or misunderstand it. It is the objects—not the individuals—that must be rescued from being categorized as damaged or deficient, and the individuals who must be taught to understand how to care for them and keep them out of the waste dump. The key words that define the subjects of *Hoarders*, in this view, bleed into their linguistic near relations: there is less difference than at first appears between hoarding, collecting, and preserving. With some sleight of hand, hoarders might actually be curators or connoisseurs, and someone's personal story about an object's value could be a story about its provenance.

All definitions of garbage attach an individual's story to larger ones about how culture defines itself, and attach a culture's stories to larger systems of production and consumption. The sheer number of popular reality television programs about how to conceptualize the value of objects—programs ranging from those that stigmatize hoarders, to those that confer the status of collector to the curators of objects—demonstrates a growing public awareness of the problem of what becomes of the commodities that seem to underwrite and constitute every iteration of subjectivity under capitalism.

The recognition that garbage is at the heart of every commodity, and thus puts the individual in tension with larger systems, is in part, we think, the reason why so many books that discuss garbage and waste, and why so many blogs and environmental debates, seem to begin by cataloging an individual household's waste. A special issue of the journal *Alphabet City* (Knechtel 2006) on trash is organized around both large- and small-scale inventories of garbage, and various explanations of its wider significance, including meditations on individual objects that have been thrown away as well as essays on dump picking. In their respective books, Heather Rogers (2005) and Elizabeth Royte (2005) give compulsively inventory of their garbage cans when they decide to follow their own trash. Philosophers Scanlan (2005) and Gay Hawkins (2006) describe items in the garbage they see around them, tracking back and forth between the material and conceptual. Popular news stories report on people and corporations attempting to attain the near-magical status of "zero waste," and discuss approvingly the day-to-day decisions about how to perform the calculus of needs and desires, repairing rather replacing, making do and doing without.

We even found ourselves doing it. As we collaboratively wrote this book, one of us anxiously tried to figure out how much drafting and editing we could do on the computer screen before we needed to print out these chapters. Later, the other got up to make a cup of tea and remembered to put the tea bag in the compost bin versus the garbage can. In general, we are confronted at every turn with the sheer amount of waste we produce in the process of daily life and the sheer amount of waste produced by corporations in order to keep our lives running. Like many other scholars of garbage, we find ourselves beginning with our own microenvironments, calculating how to reduce our personal consumption, and working, say, to get our weekly garbage down to a single bag. The energy it takes to run a computer, the final destination of laptop parts, and the energy it took to make those components—all of it seems daunting to count, catalog, narrate, and define, and all seems completely and simultaneously about both individual choices and the systems that create and naturalize them. We have no choice but to conceptualize ourselves in relationship to our stuff along with how it is produced and where it will go.

In constructing an active and creative response to these paradoxes, we—like the scholars of garbage on whom we rely as well as the authors in this collection (and we hope the readers of this volume)—find in the act of scrutinizing garbage in its incarnation as discarded objects as well as its existence as part of a larger system of consumption and production the grounds to think harder about the narratives we take for granted about the systems that sustain our culture's fantasy of the good life. In particular, we hope that our readers see in our contributors' chapters a means to conceptualize garbage and waste beyond the simple binaries that produce immediate feelings of moralizing or despair, or that ask us to see it as a problem that one human actor can confront as an individual. The very act of narrating, defining, or theorizing garbage, our contributors argue, is an act that takes seriously the transformative power of inquiry to rearrange the way we think about cause and effect, good and bad choices, and the relationship of individuals to various communities. Our contributors together contend that it is not just how we narrate garbage that matters; it's how our narratives identify the conflicts in seemingly coherent, taken-for-granted definitions.

Thus, we close this volume not with a triumphant set of solutions that reassemble core ideas to let us see what garbage and waste really are, or propose analytic methods that dramatically or definitively recast how subjects organize the material conditions and fantasies of everyday life. Neither do we purport to have identified once and for all how large corporate and political structures superintend populations by laying waste to some people and places at the expense of enriching others. We instead close by noting that each of our chapters has in some way described the utility of understanding garbage and waste as fundamentally narrative because those categories are protean, even when they are defined materially. Hearing a range of narratives from different disciplines and cultural positions can show us that our culture not only generates but also is actually built on shifting, unstable definitions of garbage and waste. Our contributors demonstrate that garbage and waste produce conversations across a range of social actors, enable debate between different interests, and are the result of varying expertise that is widely distributed among different social groupings. In this sense, the narratives we have assembled here argue that the circulation of conflicting narratives provides a way to think sociably and productively.

Each of the contributors to this volume has enacted this sociability, for together they staged a conversation about garbage from particular disciplinary vantage points, and individually, each has focused on how the idea of garbage has been defined as well as discussed at different political, historical, and cultural sites. They therefore reveal one final issue in the discussion of how we hope that the narratives about garbage have worked here, for many of our contributors have not only concentrated on the conflicting stories about garbage and waste but also on those rare moments when social actors achieve a workable consensus about them.

Environmental writers have often fallen back on an idea of community as a magical antidote to widespread ecological degradation, yet that word, like the term sustainability that seems to drive so much public debate about environmental policies, laws, regulations, and private practices, can be maddeningly vague. But as the authors here demonstrate, it is the very vagueness of that notion that can bring together the material and metaphoric aspects of garbage because it provides a provisional, temporary sense of how individual subjects unite with one another around their recognition of their similar engagement with issues like health, working conditions, pollution, toxicity, groundwater, landfills, or public works. The chapters in this volume have all shown that many of the most critical historical and cultural shifts in how to theorize garbage and waste have emerged at the moment when people recognize the limitations of individual action in the face of larger structural problems, and decide to speak as part of a community that organizes itself partly around a growing awareness that garbage and waste are not merely personal or individual problems but instead symptoms of larger crises.

This we hope is the most significant contribution of this volume. The stories our contributors have told ask us to see garbage as both material and symbolic, real and abstract, and the problem of corporations, governments, and individuals. But they also provide frameworks to understand how narrative itself can be used to retell stories we think we know in different terms. From memoirs to mapping to organized counternarratives, the chapters here propose that the very act of recognizing the interconnectedness and conflicted nature of stories about waste and garbage turns them into the site of productive action.

And as each chapter has made this argument—tacking back and forth between big issues and small, objects and systems, narratives and counternarratives as well as maps and portraits—together they show, we believe, the productiveness of staging a conversation between different disciplines. In the academic world, the way that knowledge is organized in different fields and disciplines is narrative in a deep sense; knowledge follows conventions, works according to known ways of building a story through accepted versions of evidence, and addresses readers who are accustomed to the rules of those narratives. Yet in an academic world in which there is a great deal of lip service paid to interdisciplinarity even while there is a powerful incentive to work within disciplinary boundaries, sometimes conversation fails. We cannot hear one another's stories because we do not know their conventions.

It is our hope that in bringing together scholars from different disciplines to talk about a common interest, we helped to stage such a conversation. We did not wish to create a single narrative about garbage and waste, or conversely, underscore the incommensurability of certain ways of narrating those terms. Rather, we asked scholars to address one another without giving up their disciplinary investments, and thus find a way not only to tell their stories about garbage and waste but also to tell the story of how and why their stories matter. We trust that this conversation has proved fruitful for readers.

References

Hawkins, Gay. 2006. *The Ethics of Waste: How We Relate to Rubbish.* Lanham, MD: Rowman and Littlefield.

Knechtel, John. 2006. *Trash.* Cambridge, MA: MIT Press.

Rogers, Heather. 2005. *Gone Tomorrow: The Hidden Life of Garbage.* New York: New Press.

Royte, Elizabeth. 2005. *Garbageland: The Secret Trail of Trash.* New York: Little, Brown and Company.

Scanlan, John. 2005. *On Garbage.* London: Reaktion Books.

About the Contributors

Jennifer Clapp   is a professor and Centre for International Governance Innovation chair in global environmental governance in the Department of Environment and Resources Studies at the University of Waterloo. She is the coeditor of *Global Environmental Politics* as well as serving on the editorial boards of *Global Governance* and *Alternatives Journal*. Clapp is the also the coeditor of numerous anthologies on environmental issues, including *Corporate Power in Global Agrifood Governance* (with Doris Fuchs, 2009) and *Corporate Accountability and Sustainable Development* (with Peter Utting, 2008), and the author of *Hunger in the Balance: The New Politics of International Food Aid* (2012), *Food* (2012), *Paths to a Green World: The Political Economy of the Global Environment* (cowritten with Peter Dauvergne, 2005, 2011) and *Toxic Exports: The Transfer of Hazardous Wastes from Rich to Poor Countries* (2001).

Marisol Cortez   is active in environmental justice movements, as both a scholar and an organizer. As a research assistant and preparing reports for the Environmental Justice Project out of the John Muir Center of the Environment at the University of California at Davis, she helped develop an inventory of environmental justice research needs in California's Central Valley, and she has organized around water, climate, and protection of indigenous sacred spaces in San Antonio, Texas, and Lawrence, Kansas. She is currently the American Council of Learned Societies New Faculty Fellow at the University of Kansas, where she writes and teaches in the American Studies department about the intersections of waste, embodiment, culture, nature, and power.

Stephanie Foote   is an associate professor of English as well as gender and women's studies at the University of Illinois. The recipient of a National Endowment for the Humanities Fellowship from the Winterthur Library and an Andrew Mellon Fellowship, she is the author of the book *Regional Fictions* (2001) and is completing a book titled *Signs Taken for Blunders* on class mobility in late nineteenth-century US culture. Foote is also currently writing a book on the environmental impact of the book trade.

Scott Frickel   is an associate professor of sociology at Washington State University. His central research interest lies in the study of knowledge, particularly as it relates to environmental health, risk, and justice. Frickel is the author of *Chemical Consequences: Environmental Mutagens, Scientist Activism, and the Rise of*

Genetic Toxicology (2004), which won the American Sociology Association's Robert K. Merton Book Award, and with Kelly Moore is the coeditor of *The New Political Sociology of Science: Institutions, Networks, and Power* (2006).

William Gleason   is professor of English at Princeton University, where he is also Acting Director of the Program in American Studies and Affiliated Faculty with the Princeton Environmental Institute and the Center for African American Studies. Author of *The Leisure Ethic: Work and Play in American Literature, 1840–1940* (1999) and *Sites Unseen: Architecture, Race, and American Literature* (2011), Gleason is also coeditor of the forthcoming volume, *Keywords in the Study of Environment and Culture*.

Elizabeth Mazzolini   is a visiting assistant professor of English at Virginia Tech University. She teaches courses on the intersections between science, nature, and culture, and has published reviews and essays in *Battleground: Science and Technology*, *Communication Review*, *Theory and Event*, and *Cultural Critique*. Mazzolini is completing a book about the discourse and technology surrounding Mount Everest, titled *Human Nature: Mount Everest and Material Ideology*.

Richard S. Newman   is a professor of history at the Rochester Institute of Technology. He is the author of *Freedom's Prophet: Bishop Richard Allen, the AME Church, and the Black Founding Fathers* (2008) and *The Transformation of American Abolitionism: Fighting Slavery in the Early Republic* (2002), the latter of which was a finalist for the Avery O. Craven Award of the Organization of American Historians. Along with Daniel Payne, he coedited *The Palgrave Environmental Reader* (2005), and is currently working a book-length project titled *"These 16 Acres": Love Canal and the American Dream: An Environmental History of the Love Canal Disaster*.

Phaedra C. Pezzullo   is an associate professor in the Department of Communication and Culture, and an adjunct faculty member of the Department of American Studies and Program in Cultural Studies at Indiana University, Bloomington, U.S. She is the author of *Toxic Tourism: Rhetorics of Travel, Pollution, and Environmental Justice* (2007), which won four book awards, including the Christine L. Oravec Research Award in Environmental Communication. Pezzullo also coedited *Environmental Justice and Environmentalism: The Social Justice Challenge to the Environmental Movement* (2007) and edited *Cultural Studies and the Environment, Revisited* (2010). She served as a commissioner of the City of Bloomington's Environmental Commission from 2004 to 2012 and launched the Bloomington, Indiana, PCB Oral History Project in 2012.

Daniel Schneider   is a professor in the Department of Urban and Regional Planning, and an affiliate in the Program in Ecology, Evolution, and Conservation Biology as well as Department of Entomology at the University of Illinois at Urbana-Champaign. He also currently serves as a professional scientist for the Illinois Natural History Survey. His book *Hybrid Nature: Sewage Treatment and the Contradictions of the Industrial Ecosystem* was recently published (2011).

Index

A&E (Arts and Entertainment Network), 254–255

Accountability, 120, 122, 126–127, 134, 140n22

Acetone-butanol, 178

Activism
  Brave Little Toaster, The, and, 246
  discrediting victims and, 129
  Don't Mess with Texas campaign and, 227
  Earth Day and, 129, 184
  enviroblogging and, 77, 79, 84, 87
  e-waste and, 126–135, 129–130
  garbage and, 2–3, 9–15
  grassroots, 15, 23, 27–37, 128, 130, 139n16
  Milorganite and, 185–188
  Mount Everest and, 153, 163
  New Orleans and, 108, 113
  object lessons and, 254
  plastic waste and, 200–209, 215–217
  privy reform and, 50–66
  proplastic Web sites and, 208–209
  toxic literature and, 15, 23–38, 41, 43, 45n11
  Waring and, 50–51, 58, 60–61, 65, 67
  women's efforts and, 24, 27–37, 42, 45n24

Adams, Samuel Hopkins, 136

Adhesives, 178

Agency for Toxic Substances and Disease Registry, 124

Agriculture, 121

Pennsylvania Board of Agriculture and, 63

sludge and, 180, 186, 190

U.S. Department of Agriculture and, 50, 58–59, 68n9, 186, 190

Waring and, 50–53, 58–59, 63, 67nn3,4, 68n9

Agriculture Street Landfill, 108

Ague, 52

Akra-Soilite, 181

Alexander M. McIver and Son, 174–175

Allied Mills, 179

Alphabet City journal, 257

American Beauty (film), 228

American Beverage Association, 216

American Chemistry Council, 207–208, 213, 218

American Pickers (TV show), 256

American Women's Home, The (Beecher and Beecher Stowe), 54, 67n7

Amyotrophic lateral sclerosis (ALS), 186, 188

Andrews, Mike, 128

Angelus Novus (Klee), 242

Animal, Vegetable, Miracle (Hopp and Kingsolver), 86

Animal Hoarders (TV show), 255

Anthrax, 179

Anthropology, 8, 10, 233

Antiques Roadshow (TV show), 256

Antonetta, Susan, 38–39

Apa (tour guide), 158–161

Urban and Industrial Environments

Series editor: Robert Gottlieb, Henry R. Luce Professor of Urban and Environmental Policy, Occidental College

Eran Ben-Joseph, *The Code of the City: Standards and the Hidden Language of Place Making*

Nancy J. Myers and Carolyn Raffensperger, eds., *Precautionary Tools for Reshaping Environmental Policy*

Kelly Sims Gallagher, *China Shifts Gears: Automakers, Oil, Pollution, and Development*

Kerry H. Whiteside, *Precautionary Politics: Principle and Practice in Confronting Environmental Risk*

Ronald Sandler and Phaedra C. Pezzullo, eds., *Environmental Justice and Environmentalism: The Social Justice Challenge to the Environmental Movement*

Julie Sze, *Noxious New York: The Racial Politics of Urban Health and Environmental Justice*

Robert D. Bullard, ed., *Growing Smarter: Achieving Livable Communities, Environmental Justice, and Regional Equity*

Ann Rappaport and Sarah Hammond Creighton, *Degrees That Matter: Climate Change and the University*

Michael Egan, *Barry Commoner and the Science of Survival: The Remaking of American Environmentalism*

David J. Hess, *Alternative Pathways in Science and Industry: Activism, Innovation, and the Environment in an Era of Globalization*

Peter F. Cannavò, *The Working Landscape: Founding, Preservation, and the Politics of Place*

Paul Stanton Kibel, ed., *Rivertown: Rethinking Urban Rivers*

Kevin P. Gallagher and Lyuba Zarsky, *The Enclave Economy: Foreign Investment and Sustainable Development in Mexico's Silicon Valley*

David N. Pellow, *Resisting Global Toxics: Transnational Movements for Environmental Justice*

Robert Gottlieb, *Reinventing Los Angeles: Nature and Community in the Global City*

David V. Carruthers, ed., *Environmental Justice in Latin America: Problems, Promise, and Practice*

Tom Angotti, *New York for Sale: Community Planning Confronts Global Real Estate*

Paloma Pavel, ed., *Breakthrough Communities: Sustainability and Justice in the Next American Metropolis*

Anastasia Loukaitou-Sideris and Renia Ehrenfeucht, *Sidewalks: Conflict and Negotiation over Public Space*

David J. Hess, *Localist Movements in a Global Economy: Sustainability, Justice, and Urban Development in the United States*

Julian Agyeman and Yelena Ogneva-Himmelberger, eds., *Environmental Justice and Sustainability in the Former Soviet Union*

Jason Corburn, *Toward the Healthy City: People, Places, and the Politics of Urban Planning*

JoAnn Carmin and Julian Agyeman, eds., *Environmental Inequalities Beyond Borders: Local Perspectives on Global Injustices*

Louise Mozingo, *Pastoral Capitalism: A History of Suburban Corporate Landscapes*

Gwen Ottinger and Benjamin Cohen, eds., *Technoscience and Environmental Justice: Expert Cultures in a Grassroots Movement*

Samantha MacBride, *Recycling Reconsidered: The Present Failure and Future Promise of Environmental Action in the United States*

Andrew Karvonen, *Politics of Urban Runoff: Nature, Technology, and the Sustainable City*

Daniel Schneider, *Hybrid Nature: Sewage Treatment and the Contradictions of the Industrial Ecosystem*

Catherine Tumber, *Small, Gritty, and Green: The Promise of America's Smaller Industrial Cities in a Low-Carbon World*

Sam Bass Warner and Andrew H. Whittemore, *American Urban Form: A Representative History*

John Pucher and Ralph Buehler, eds., *City Cycling*

Stephanie Foote and Elizabeth Mazzolini, eds., *Histories of the Dustheap: Waste, Material Cultures, and Social Justice*